Modelling, Forecasting, Artificial Neural Network and Expert System in Fisheries and Aquaculture

MODELLING FORECASTING ARTIFICIAL NEURAL NETWORK AND EXPERT SYSTEM IN FISHERIES AND AQUACULTURE

Ajit Kumar Roy
Niranjan Sarangi

2026

Daya Publishing House®
A Division of
Astral International Pvt. Ltd.
New Delhi – 110 002

First Published, 2009
Reprinted, 2026

ISBN: 978-81-7035-571-7

Publisher's Note:

Every possible effort has been made to ensure that the information contained in this book is accurate at the time of going to press, and the publisher and author cannot accept responsibility for any errors or omissions, however caused. No responsibility for loss or damage occasioned to any person acting, or refraining from action, as a result of the material in this publication can be accepted by the editor, the publisher or the author. The Publisher is not associated with any product or vendor mentioned in the book. The contents of this work are intended to further general scientific research, understanding and discussion only. Readers should consult with a specialist where appropriate.

Every effort has been made to trace the owners of copyright material used in this book, if any. The author and the publisher will be grateful for any omission brought to their notice for acknowledgement in the future editions of the book.

Published by : **Daya Publishing House®**
A Division of
Astral International Pvt. Ltd.
– ISO 9001:2015 Certified Company –
4736/23, Ansari Road, Darya Ganj
New Delhi-110 002
Ph. 011-43549197, 8130496929
E-mail: info@astralint.com
Website: www.astralint.com

Printed at : **Replika Press Pvt. Ltd.**

डा. एस अय्यप्पन
उपमहानिदेशक (मत्स्य)
DR. S. AYYAPPAN
Deputy Director General
(Fisheries)

ICAR

भारतीय कृषि अनुसंधान परिषद
कृषि अनुसंधान भवन-२,
पूसा, नई दिल्ली - 110 012
INDIAN COUNCIL OF
AGRICULTURAL RESEARCH
KRISHI ANUSANDHAN BHAVAN-II,
PUSA, NEW DELHI – 110 012

Foreword

With a bountiful biophysical resource base India is well-positioned to be a more ignificant player of aquaculture production as it has established itself as one of the astest growing food producing sector. It contributes about 4.2% of global aquaculture roduction of 45.5 million tones. Apart from supplying cheap protein, this sector is ontributing significantly in poverty alleviation, food security and income generation or rural population. Demand from aquaculture sector is increasing because roduction from capture fisheries has leveled off. In recent past Government of India as given due attention for all round development of aquaculture sector through stablishment of National Fisheries Development Board to augment the productivity urther on a sustained basis, through capacity building in the form of training of nanpower in adoption and application of viable technologies across the country. To upport the government initiative reliable, accurate and timely statistics on future lemand and supply is needed. Under the situation prediction and forecasting of ikely population, production, productiveness of different component of fish roduction system involving energy and water requirement, whether, climate disease nd paste, price and market demand and supply for different fishery production is ssential for planners.

I am extremely happy to learn that a book entitled 'Modelling, Forecasting, rtificial Neural Network and Expert System in Fisheries and Aquaculture' has been ompiled and edited with up-to-date and interesting articles contributed by the minent professionals of the national institutes. Application of various forecasting

Ph: (O) 25846738, (R) 25843190; Fax: 00-91-11-25841955
E-mail: ayyappans@icar1.nic.in & s_ayyappans @yahoo.com

methods provides an accessible and comprehensive account of recent developments in statistical and economic forecasting as this enables unraveling of underlying relationship among crucial variables in fisheries research. Each of the sixteen chapters of this book contributed by experts working in the field deals not only with methodological aspects of estimation of fish stock, modelling, forecasting, neural network, fuzzy logic, soft computing, productivity analysis, quantitative and qualitative analysis but also application on real data with illustrated examples using statistical and econometric packages. Some of the chapters will give insight for planners about the future scenario of fish production which in turn help taking appropriate decisions.

It is expected that this publication will be useful for research workers and planners engaged in fisheries research equally that may serve as a unique source of reference material on the application latest tools and techniques on modelling and forecasting.

I wish this maiden endeavor a grand success.

S. Ayyappan

Preface

Most challenging task to the planners of fisheries and aquaculture is to produce food and nutritional security to the present population as well as to plan for supply of food to the future population. Therefore, planners should have clear idea about the possible future population and accordingly their food and nutrition requirement. Prediction and forecasting of likely population, production, likely prediction of productiveness of different component of fish production system. Forecasting of energy and water requirement, whether, climate, disease and paste, price and market demand and supply for different fishery production is essential for sustainable fish production.

Keeping these points in view the book has been edited that provides an accessible and comprehensive account of recent developments in modelling and forecasting in fisheries and aquaculture. Each chapter has been specially written by an expert in the field, bringing together in a single volume a range of approaches and views. Uniquely, modelling and forecasting in a single volume, the book provides a comprehensive account of the leading approaches and modelling strategies that are routinely used. This book is basically the result of twelve years of teaching the material to state officials and students during conducting yearly workshops. This book further aims to bring the principles, techniques for modelling and forecasting of fish production for both capture and culture fisheries.

The book contains the following sixteen chapters namely 1.Statistical modelling in Fisheries Research: An Overview 2. Is Fisheries Sector in India Changing its Gear? 3. Non–Linear Models for Forecasting Fish Production from Ponds 4. Statistical Models and their Application in Fishery Science Research 5. Methods of Forecasting and their Application in Fishery Science 6. Modelling in Aquaculture 7. Modelling of Multiobjective Freshwater Aquaculture 8. A Revisit to Building up Models in Fishery Research 9. Quantitative Techniques for Fish Stock Assessment 10. Time Series

Modelling and its Applications in Fisheries 11. Tropho-Dynamic Modelling for Ecosystem Based Fisheries Management 12. Forecasting of Fish Production of India by culture environment and species 13. Expert Systems and Simulation Studies in Agriculture/Aquaculture 14. Recent Trends in Artificial Neural Network (ANN) and Fuzzy Logic and their Application in Aquaculture 15. Soft Computing Techniques for Data Analysis and Modelling in Fisheries 16. An Econometric Model for Productivity Analysis of Aquaculture Farms.

The publication is likely to cater to the need of a wide audience of students, researchers, scientists, professionals, policy makers and in fact anyone who works in aquaculture and fisheries.

The material owes its existence to many conversations and collaborations with many resource persons delivering invited talks. I am very grateful to them for their patience and tolerance as it progressed from crude notes to its present form. The endeavor could not have been materialized without the active support in terms of contributing articles by the dedicated subject matter specialists, professionals from various National Institutes of repute. All out support, guidance, suggestion and encouragements from Dr. M.K. Bhan, Secretary, DBT and Dr. T.Madhan Mohan, Advisor, DBT are thankfully acknowledged. Heart felt thanks are due to Dr. S. Ayyappan, DDG (Fy.), ICAR for constant encouragement to make the programme a success.

Invitation to resource persons, conducting the workshop and subsequent publication of this book would not have been possible without the financial support of Bioinformatics Division of Dept. of Biotechnology, Ministry of Science & Technology, and Government of India.

Ajit Kumar Roy

Niranjan Sarangi

Contents

List of Contributors

1. **Statistical Modelling in Fisheries Research: An Overview**
Prajneshu
Head, Biometrics Division
Indian Agricultural Statistics Research Institute,
Library Avenue, New Delhi – 110 012

2. **Is Fisheries Sector in India Changing its Gear?**
B. Ganesh Kumar
Senior Scientist
National Centre for Agricultural Economics & Policy Research,
P.B.No.11305, DPS Marg, Pusa, New Delhi – 110 012

3. **Non–Linear Models for Forecasting Fish Production from Ponds**
Lalmohan Bhar
Senior Scientist,
Indian Agricultural Statistics Research Institute,
Library Avenue, New Delhi – 110 012

4. **Statistical Models and their Application in Fishery Science Research**
R. Venugopalan
Senior Scientist
Agricultural Statistics, Indian Institute of Horticultural Research (ICAR),
Bangalore – 560 089

Prajneshu
Head, Biometrics Division
Indian Agricultural Statistics Research Institute,
Library Avenue, New Delhi – 110 012

5. **Methods of Forecasting and their Application in Fishery Science**

P.K. Sahu

Reader
Department of Agricultural Statistics,
Bidhan Chandra Krishi Viswa Vidyalaya,
Mohanpur, Nadia, West Bengal

6. **Modelling in Aquaculture**

A.K. Roy

Principal Scientist & Head,
Social Science Section, Central Institute of Freshwater Aquaculture (CIFA),
Bhubaneswar – 751 002

7. **Modelling of Multiobjective Freshwater Aquaculture**

Pratap K.J. Mohapatra

Head,
Industrial Engineering Department,
Indian Institute of Technology,
Kharagpur – 721 302

Santosh K. Prusty

Research Fellow
Indian Institute of Technology, Kharagpur – 721 302

C. K. Mukherjee

Professor,
Department of Agriculture & Food Engineering
Indian Institute of Technology, Kharagpur – 721 302

8. **A Revisit to Building up Models in Fishery Research**

Satyabrata Pal

Ex-Dean,
Bidhan Chandra Krishi Viswa Vidyalaya
Mohanpur, Nadia, West Bengal

Subhabaha Pal

Researcher,
Bidhan Chandra Krishi Viswa Vidyalaya
Mohanpur, Nadia, West Bengal

9. **Quantitative Techniques for Fish Stock Assessment**

R.S. Biradar

Principal Scientist & Head,
Central Institute of Fisheries Education,
Deemed University, Fishery University Road,
Seven Bungalows, Versova,
Mumbai – 400 061

10. Time Series Modelling and its Applications in Fisheries

S. Ravichandran

Sr. Scientist (Statistics),
Directorate of Rice Research,
Rajendranagar, Hyderabad - 500 030

11. Tropho-Dynamic Modelling for Ecosystem Based Fisheries Management

Preetha Panikkar

Scientist,
Reservoir Division, Central Inland Fisheries Research Institute,
Hessarghatta Lake Post,
Bangalore – 560 089

M. Feroz Khan

Sr. Scientist,
Reservoir Division, Central Inland Fisheries Research Institute,
Hessarghatta Lake Post,
Bangalore – 560 089

12. Forecasting of Fish Production of India by Culture Environment and Species

A. K. Roy

Principal Scientist & Head,
Social Science Section, Central Institute of Freshwater Aquaculture (CIFA),
Bhubaneswar – 751 002

Nirupama Panda

Technical Officer (T-6),
Social Science Section, Central Institute of Freshwater Aquaculture (CIFA),
Bhubaneswar – 751 002

13. Expert Systems and Simulation Studies in Agriculture/Aquaculture

J. Panduranga Rao

Consultant,
Agricultural Finance Corporation,
Sirisuban Apartments, Somajigoda
Hyderabad

14. Recent Trends in Artificial Neural Network (Ann) and Fuzzy Logic and their Application in Aquaculture

M. Balakrishnan

Co-ordinator, BTIS,
Central Agricultural Research Institute,
Port Blair – 744 101, Andaman & Nicobar Islands

R.C. Srivastava

Director,
Central Agricultural Research Institute,
Port Blair – 744 101, Andaman & Nicobar Islands

15. Soft Computing Techniques for Data Analysis and Modelling in Fisheries

H.C. Verma

Scientist (SS),
Water Technology Centre for Eastern Region (ICAR),
Chandrasekharpur, Bhubaneswar – 751 023

16. An Econometric Model for Productivity Analysis of Aquaculture Farms

A.K. Roy

Principal Scientist & Head,
Social Science Section, Central Institute of Freshwater Aquaculture (CIFA),
Bhubaneswar – 751 002

G. S. Saha

Sr. Scientist,
Social Science Section, Central Institute of Freshwater Aquaculture (CIFA),
Bhubaneswar – 751 002

N. Sarangi

Director,
Central Institute of Freshwater Aquaculture (CIFA),
Bhubaneswar – 751 002

Nibedita Jena

Economic Investigator,
Directorate of Economic Research,
Khadi & Village Industry Commission,
Gramodaya, 3, Irla Road, Vile Parle (West),
Mumbai – 400 056

Modelling, Forecasting, Artificial Neural Network and Expert System in Fisheries
Editors: Ajit Kumar Roy & Niranjan Sarangi
Published by: DAYA PUBLISHING HOUSE (2009) *Pages* 1–12

Chapter 1

Statistical Modelling in Fisheries Research: An Overview

Prajneshu

Introduction

Statistical modelling enables unraveling of underlying relationships among crucial variables in fisheries research. A statistical model is generally represented by an equation or a set of equations for describing the behaviour of a system. Some examples are: Length-weight relationship, Age-length relationship, Input-output relationship, Computation of compound growth rate, Catch-effort relationship, and Production over time. A statistical model can be either 'linear' or 'nonlinear'. A linear model is one in which all the parameters appear linearly. Some examples of linear model are:

Multiple Linear Regression Model

If Y is response variable, X_i are predictors, and ε is error term, then the model is

$$Y = a_0 + a_1 X_1 + \ldots + a_p X_p + \varepsilon,$$

Polynomial Models

$$Y = a + b X + \varepsilon \qquad \text{(First-order model)}$$

$$Y = a + b X + c X^2 + \varepsilon \qquad \text{(Second-order or Quadratic or Curvilinear model).}$$

'Method of least squares' is generally employed for estimation of parameters of above models.

It is well recognized that any type of statistical inquiry in which principles from some body of knowledge enter seriously into the analysis is likely to lead to a 'Nonlinear model' (Seber and Wild, 2003). A 'nonlinear model' is one in which at least one of the parameters appears nonlinearly. More formally, in a "nonlinear model", at least one derivative with respect to at least one parameter involves that parameter. Examples of a "nonlinear model" are:

$$Y(t) = exp(at+bt^2) \tag{1a}$$

$$Y(t) = at + exp(-bt). \tag{1b}$$

Note: Some authors use the term 'intrinsically linear' to indicate a nonlinear model, which can be transformed to a linear model by means of some transformation. For example, the model given by eq.(1a) is 'intrinsically linear' in view of the transformation $X(t) = log_e Y(t)$.

Some Important Nonlinear Growth Models

These models describe growth behaviour over time. The type of model needed in a specific situation depends on type of growth that occurs. In general, growth models are 'Mechanistic' in nature, rather than 'Empirical', like Multiple regression model, and Polynomial model. In the former, parameters have meaningful biological interpretation; the latter is just like a 'black-box' where some input is given and some output is taken out. A mechanistic model usually arises as a result of making assumptions about the type of growth, writing down differential or difference equations that represent these assumptions, and then solving these equations to obtain a growth model. The utility of such models is that, on one hand, they help us to gain insight into the underlying mechanism of the system and, on the other hand, they are of immense help in efficient fisheries management. We now discuss briefly some well-known nonlinear growth models:

Malthus Model

If $N(t)$ denotes the population size or biomass at time t and r is intrinsic growth rate, then rate of growth of population size is given by

$$dN/dt = rN \tag{2}$$

Integrating, we get

$$N(t) = N_0\, exp\,(rt) \tag{3}$$

where, N_0 denotes the population size at $t=0$. Thus this law entails an exponential increase for $r>0$. Furthermore, $N(t) \rightarrow \infty$ as $t \rightarrow \infty$, which can not happen in reality.

Note: In subsequent discussion, parameter r is assumed to be positive.

Monomolecular Model

This model describes the progress of a growth situation in which it is believed that the rate of growth at any time is proportional to the resources yet to be achieved, *i.e.*

$$dN/dt = r(K-N) \tag{4}$$

where, K is carrying size of the system. Integrating eq.(4), we get

$$N(t) = K - (K-N_0)\, exp\,(-rt) \tag{5}$$

The graph of $N(t)$ versus t is monotonically increasing and ultimately reaches K.

Logistic Model

This model is represented by the differential equation

$$dN/dt = rN\,(1-N/K) \tag{6}$$

Integrating, we get

$$N(t) = K/[1+(K/N_0-1)\,exp(-rt)] \tag{7}$$

The graph of $N(t)$ versus t is elongated S-shaped and the curve is symmetrical about its point of inflexion.

Gompertz Model

This is another model having a sigmoid type of behaviour and is found to be quite useful in biological work. However, unlike the logistic model, this is not symmetric about its point of inflexion. The differential equation for this model is

$$dN/dt = rN\,log_e\,(K/N) \tag{8}$$

Integration of this equation yields

$$N(t) = K\,exp[log_e\,(N_0/K)\,exp(-rt)]. \tag{9}$$

Richards Model

This model is given by

$$dN/dt = rN\,(K^m - N^m)/(mK^m) \tag{10}$$

which, on integration, gives

$$N(t) = KN_0/[N_0 + (K^m - N_0^m)\,exp\,(-rt)]^{1/m} \tag{11}$$

Evidently, the last three models are particular cases of this model when $m= -1,1,0$ respectively. However, unlike the earlier models, this model has an additional parameter.

Allometric Model

If the relative growth rates of two variables X and Y, given respectively as $X^{-1}\,dX/dt$ and $Y^{-1}\,dY/dt$ experience throughout development a constant ratio, differential equation form of allometric model is:

$$Y^{-1}\,dY/dt = bX^{-1}\,dX/dt \tag{12}$$

which, on integration, gives

$$Y = aX^b \tag{13}$$

Cobb–Douglas Production Function Model

If $X_1, X_2,..., X_p$ denote various inputs and Y is corresponding output, the model is given by:

$$Y = a X_1^{b_1} X_2^{b_2} \dots X_p^{b_p} \tag{14}$$

Fitting of Nonlinear Models

The above models have been posed deterministically. Obviously, this is unrealistic and so we replace these deterministic models by statistical models by adding an error term on the right hand side and making appropriate assumptions about them. This results in a 'Nonlinear statistical model'. As in linear regression, in non-linear case also, 'Method of least squares' can be employed to obtain parameter estimates. However, minimization of residual sum of squares yield normal equations, which are nonlinear in parameters. Since it is not possible to solve nonlinear equations exactly, next alternative is to obtain approximate analytical solutions by employing iterative procedures. Three main methods of this kind are:

1. Linearization (or Taylor Series) method
2. Steepest descent method
3. Levenberg-Marquardt's method

Details of above methods along with their merits and demerits are given in Draper and Smith (1998). Linearization method uses the results of linear least square theory in a succession of stages. However, neither this method nor Steepest descent method, is ideal. The latter method is able to converge on true parameter values even though initial trial values are far from true parameter values, but this convergence tends to be very slow at latter stages of iterative process. On the other hand, Linearization method will converge very rapidly provided vicinity of true parameter values has been reached, but if initial trial values are too far removed, convergence may not occur at all.

Most widely used method of computing nonlinear least squares estimates is Levenberg-Marquardt's method. This method represents a compromise between the other two methods; it combines successfully best features of both, and avoids their serious disadvantages. It is good in the sense that it usually converges and does not 'slow down' at latter part of the iterative process. We now discuss this method in some detail.

Let us consider the model

$$y_i = f(x_i, \theta) + \varepsilon_i, i = 1,2,\dots,n, \tag{15}$$

where, (x_i, y_i) denote respectively the ith observations of predictor and response variables, $\boldsymbol{\theta} = (\theta_1 \ \theta_2 \dots \theta_p)'$ are parameters, and ε_i, the error terms are assumed to be independent and identically distributed as $N(0, \sigma^2)$. The residual sum of squares is

$$S(\boldsymbol{\theta}) = \sum_{i=1}^{n} [y_i - f(x_i, \boldsymbol{\theta})]^2. \tag{16}$$

Let $\boldsymbol{\theta}_0 = (\theta_{10}\ \theta_{20} \ldots \theta_{po})'$ be the vector of initial parameter values. Then the algorithm for obtaining successive estimates is essentially given by

$$(H + \tau\, I)(\boldsymbol{\theta}_0 - \boldsymbol{\theta}_1) = \mathbf{g}, \tag{17}$$

where,

$$\mathbf{g} = \left.\frac{\partial S(\boldsymbol{\theta})}{\partial \boldsymbol{\theta}}\right|_{\boldsymbol{\theta}=\boldsymbol{\theta}_0}, \quad H = \left.\frac{\partial^2 S(\boldsymbol{\theta})}{\partial \boldsymbol{\theta}\, \partial \boldsymbol{\theta}'}\right|_{\boldsymbol{\theta}=\boldsymbol{\theta}_0},$$

I is the identity matrix and τ is a suitable multiplier.

Notes

(*i*) Marquardt scaled various quantities appearing in (17) by standard deviations of derivatives with respect to parameter values.

(*ii*) Now a days, most of the standard statistical packages contain computer programs to fit nonlinear statistical models based on Levenberg-Marquardt algorithm. For example, SPSS has NLR option, SAS has NLIN option, IMSL has RNSSQ option to accomplish the task.

Choice of Initial Values

All procedures for nonlinear estimation require initial values of parameters and choice of good initial values is very crucial. However, there is no standard procedure for getting initial estimates. Most obvious method for making initial guesses is by use of prior information. Estimates calculated from previous experiments, known values for similar systems, values computed from theoretical considerations yield ideal initial guesses. Some other methods are:

(*i*) *Linearization*. After ignoring error term, check form of the model to see if it could be transformed into a linear form by means of some transformation. In such cases, linear regression can be used to obtain initial values.

(*ii*) *Solving a system of equations*. If there are p parameters, substitute for p sets of observations in the model ignoring error. Solve these equations for parameters, if possible. Widely separated x_i often work best.

(*iii*) *Using properties of the model*. Consider behaviour of response function as x_i go to zero or infinity, and substitute for observations that most nearly represent those conditions in scale and context of the problem; solve, if possible, resulting equations.

(*iv*) *Graphical method*. Sometimes a visual estimate can be obtained by plotting data.

Goodness of Fit of a Nonlinear Model

This is generally assessed by Coefficient of determination, R^2. However, as pointed out by Kvalseth(1985), eight different expressions for R^2 appear in the literature.

Huang and Draper (2003) have suggested yet another measure based on R^2. One of the most frequent mistakes occurs when fits of a linear and a nonlinear model are compared by using the same R^2 expression but different variables. Thus, for example, a power model or an exponential model may first be linearized by using a logarithmic transformation and then fitted to data by using ordinary least squares method. The R^2-value is then often calculated using data points $(log_e\ y_i,\ log_e\ \hat{y}_i)$. The R^2 is generally interpreted as a measure of goodness of fit of even the original nonlinear model, which is incorrect. Scott and Wild (1991) have given a real example where two models are identical for all practical purposes but have very different values of R^2 calculated on transformed scales. Kvalseth (1985) has emphasized that, although R_1^2 given by

$$R_1^2 = 1 - \Sigma(y_i - \hat{y}_i)^2 / \Sigma(y_i - \bar{y})^2 \tag{18}$$

is quite appropriate even for nonlinear models, other summary statistics, like

Mean Absolute Error (MAE) = $\Sigma\left|y_i - \hat{y}_i\right| / n,$

Mean Squared Error (MSE) = $\Sigma(y_i - \hat{y}_i)^2 / (n-p),$

should also be computed. Here n is total number of observed values and p denotes number of model parameters.

Examination of Residuals

Uncritical use and sole reliance on above statistics may fail to reveal important data characteristics and model inadequacies. Additional detailed analysis of residuals is strongly recommended to decide about suitability of a model. Two important assumptions made in the model are:

(*i*) Errors are independent

(*ii*) Errors are normally distributed.

Assumption of independence of errors is required for estimation of parameters while assumption of normality is required only for hypothesis testing. These assumptions can be verified by examining residuals. If fitted model is correct, the residuals should exhibit tendencies that tend to confirm or at least should not exhibit a denial of the assumptions. Well-known Run test may be used for (i), while Shapiro-Wilk test may be employed for examining assumption (ii).

Growth Models in Fisheries

A "National freshwater aquaculture development plan" has been formulated (Gopakumar *et al.*, 1999). Growth models are widely used in fisheries and a detailed review is given in Prajneshu (2005). Some of these models are discussed below:

Length–weight Relationship

Length (L)–weight (W) relationship is generally determined by applying well-known allometric model given by eq.(13). Usual practice is to first make logarithmic transformation and then add an error term (ε), thus getting

$$log W = log\, a + b\, log\, L + \varepsilon \tag{19}$$

The advantage of doing this is that, in terms of variables *logW* and *logL*, eq.(19) is a linear equation, and so its parameters can easily be estimated by applying Method of least squares. However, it may be noted that eqs.(13) and (19) are identical only when a multiplicative error term is assumed in the original eq.(13). This assumption tends to be valid only when variability in response variable (W) increases as W increases. However, proper justification for this assumption is rarely provided and this choice is generally made only for mathematical convenience. Another mistake generally committed is to assess goodness of fit of original model given by eq.(13) by taking the same value of coefficient of determination R^2 as that for transformed model given by eq.(19). Further, estimates of standard errors of original parameters, *viz.* a and b can not be computed. In view of all this, it is desirable that eq.(13) with additive error should be fitted to data using procedures of nonlinear estimation, as discussed by Venugopalan and Prajneshu (1998).

As, generally, no attention is paid to reparameterization of parameters, model-data set exhibits an extremely high nonlinear behaviour. To this end, Prajneshu and Ravichandran (2001) used "Expected-value parameters", which depict close-to-linear behaviour. Venugopalan and Prajneshu (1997a) studied a generalization of allometric model through the model

$$log W = log\, a + b\left(log\, L\right)^c + \varepsilon \tag{20}$$

where, ε is the error term. For $c = 1$, eq.(18) reduces to eq.(19). Wald's test may be used for testing $H_0 : c = l$ against $H_0 : c \neq l$. As an illustration, methodology was applied to determine length-weight relationship of Indian pearl oyster. Another direction in which allometric model can be extended is to consider more realistic situation that underlying phenomenon is 'fuzzy' or 'vague' instead of assuming it to be crisp. To this end, Kandala and Prajneshu (2004b) fitted allometric model by using concept of fuzzy least squares and demonstrated its superiority over traditional least squares approach.

Cobb-Douglas Production Function

As pointed out by McCausland *et al.* (2006), Cobb-Douglas production function is quite often used in fisheries to determine input-output relationships. Usual practice of fitting this model is to adopt short cut of first linearizing the model and then using method of least squares for estimation of parameters. As discussed above, such a procedure has severe drawbacks and so the model should be fitted by applying "Nonlinear estimation procedures".

Age-length Relationship

von Bertalanffy growth model (VBGM), discussed in eq.(5), is generally employed for obtaining age (t)-length(L_t) relationship. It may be expressed as

$$L_t = L_\infty \{1-exp[-k(t-t_0)]\} + \varepsilon \tag{21}$$

where, L_∞, k and t_0 are parameters. A heartening feature of this nonlinear model is that it becomes "Linear", when L_{t+1} is expressed in terms of L_t, *i.e.*

$$L_{t+1} = a+b\, L_t \tag{22}$$

where,

$$a = L_\infty [1-exp(-k)], \text{ and } b = exp(-k)$$

Venugopalan and Prajneshu (1997b) discussed nonlinear estimation procedure for fitting VBGM with auto correlated errors and applied it to age-length data under assumptions of monophasic and diphasic growth.

In traditional VBGM, response variable, *viz.* length is assumed to be described by a single number for each age group of fish. However, in reality, length corresponding to even-aged fish always lies in an interval. To this end, fuzzy regression methodology is thoroughly discussed and applied by Kandala and Prajneshu (2004a) to some data culled from fisheries literature. Recently, Singh *et al.* (2007) revisited fitting of fuzzy von Bertalanffy growth model when response variable is reported in intervals corresponding to various values of predictor variable. Specifically, a more efficient two-stage procedure based on fuzzy least squares is employed and relevant computer programs are developed in "Nonlinear programming solver LINGO, Version 8" software package. Finally, an illustration to pearl oyster age-length data is discussed.

Computation of Compound Growth Rates

These are widely used in fisheries as these have important policy implications. Current procedures for its computation is discussed below:

If y_t denotes observation (*e.g.* fisheries production, productivity, or area) at time t and r is compound growth rate, model employed for estimating r is based on the equation:

$$y_t = y_0\,(1 + r)^t \tag{23}$$

Usual practice is to assume a multiplicative error term $exp\ (\varepsilon)$ in eq.(21) so that the model may be linearized by means of logarithmic transformation. Thus,

$$ln(y_t) = A + Bt + \varepsilon \tag{24}$$

where, $A = ln\ (y_0)$, and $B = ln\ (1 + r)$. Eq.(22) is then fitted to data using "Method of least squares" and goodness of fit is assessed by coefficient of determination R^2. Finally, compound growth rate is estimated as

$$\hat{r} = exp\,(\,\hat{B}\,) - 1 \tag{25}$$

The disturbing feature is that above procedure is beset with many pitfalls. Firstly, let us consider the basic model itself as given in eq.(23). Evidently, this is nothing but famous Malthusian model proposed towards the end of eighteenth century. A well-known drawback of this model is that response variable y_t tends to infinity as $t \to \infty$, which can not happen in reality. Therefore, continuing with only this model does not make much sense, particularly when several other more realistic models exist in literature. Second drawback is that proper justification for assumption of multiplicative error is hardly ever provided. The third drawback is that there is no way to compute standard errors of estimates of original parameters, *viz.* y_0 and r appearing in eq.(23) from corresponding ones of the transformed model given in eq.(24). Lastly, goodness of fit of original model given by eq.(23) is assessed by taking same value of R^2 as computed for linearized version given by eq.(24). In fact, it is not at all possible to get R^2 for eq.(23) from that for eq.(24). Furthermore, different software packages employ different formulae for computation of R^2 and hardly any effort is made to find out as to which particular formula has been employed in the software package one is using!

Evidently, compound growth rates should be computed by first identifying that model which describes satisfactorily the path followed by response variable over time. To this end, exponential model given by eq.(23) may be replaced by more realistic growth models discussed earlier. After successfully identifying best model for a data set, last step is to compute 'Annual compound growth rate', which is given as $d\,[ln\,(y)]/dt$, *i. e.* $y^{-1}\,dy/dt$. For monomolecular, logistic, and Gompertz models, annual growth rates pertaining to the period (t_i, t_{i+1}), $i = 0, 1, \ldots, n-1$, where n denotes number of data points, on using eqs.(5), (7), and (9), are respectively.

$$R_t^M = r\,[\,K / y(t) - 1\,] \tag{26}$$

$$R_t^L = r\,[\,1 - y(t) / K\,] \tag{27}$$

$$R_t^G = r\,ln[\,K / y(t)\,] \tag{28}$$

Taking arithmetic mean, requisite Compound growth rate may be obtained (Chandran and Prajneshu, 2005).

Catch-effort Relationship

Importance of surplus production models has been highlighted by Willmann (2005). These models, *viz.* Schaefer, and Fox models are obtained respectively from logistic, and Gompertz models by subtracting a term (EN) corresponding to harvesting and then obtaining equilibrium solutions. The catch (Y)-effort (E) relationships for these models are respectively:

$$Y = K\,E\left(1 - E/r\right),\ E < r \tag{29}$$

$$Y = K\,E\,exp\left(-E/r\right) \tag{30}$$

Further, maximum sustainable yield (*MSY*) for Schaefer model is $rK/4$ with optimum effort (E^*) as $r/2$ while for Fox model, $MSY = rK/e$, $e = 2.718$, and $E^* = r$. As eqs. (27) and (28) are nonlinear, these should be fitted by applying "Nonlinear estimation procedures" (Prajneshu, 1991). A more general surplus production model, *viz.* Pella-Tomlinson model can similarly be obtained from Richards model. This model when errors are autocorrelated has been applied by Venugopalan and Prajneshu (2000) to catch-effort data of Pacific Bigeye tuna with a view to determining maximum sustainable yield and corresponding optimum effort.

Fisheries Production Over Time

Statistical modelling for determining trends in landings of various fish species, primarily employing Box-Jenkins autoregressive integrated moving average (ARIMA) methodology, was carried out by Prajneshu and Venugopalan (1997). Ravichandran and Prajneshu (2001) used Structural time-series (STM) methodology, when trend is prominent, for describing marine, inland, and total fish production of the country and demonstrated its superiority over ARIMA approach by computing various measures of goodness of fit. Ravichandran and Prajneshu (2000) modelled all-India shrimp export data when series under consideration follows an autoregressive process.

Main limitation of both the above approaches is that resultant model is 'linear', which generally does not hold in reality. Fortunately, during the last two decades or so, a number of families of nonlinear time-series modelling are developed. One such important class is that of 'bilinear time-series models' (Terdik, 1999). This family may be employed for modelling and forecasting those data sets that depict bursts at irregular time epochs. Ghosh *et al.* (2006) employed this model to describe India's marine products export data during the period 1961-62 to 1998-99 obtained from Marine Products Export Development Authority (MPEDA), Kochi, India. During the first phase from 1961-62 to 1989-90, there was more or less a gradual increase in marine products export except for sudden decrease during 3 years, *viz.* 1965-66, 1981-82 and 1985-86. In the first four years of the second phase, *i.e.*1990-91 to 1993-94, the data depicts a much higher jump. Subsequently, there was a leap in 1994-95 followed by a sudden decrease in next year. The same pattern continued for next two years and finally, there was again a sudden decrease in 1998-99. A perusal shows that shocks took place during the years 1963-64, 1973-74, 1978-79, 1983-84, 1990-91, 1994-95, 1996-97 and 1998-99, which were at irregular time epochs. Further, top five positive shocks occurring during the years 1996-97, 1994-95, 1992-93, 1990-91, 1978-79 were of uniformly greater magnitude than the top five negative ones occurring during 1998-99, 1995-96, 1980-81, 1974-75, 1984-85, except the year 1998-99. Thus, in short, for India's marine products export data, shocks took place at irregular time epochs and positive shocks were generally of greater magnitude than the negative ones. Under this scenario, it is more appropriate to study bilinear time-series models, which are capable of taking into account the effect of past observations with random coefficients, which are functions of past errors in a multiplicative manner, unlike the ARIMA family. The best model from the family of bilinear time-series models was identified on the basis of minimum normalized Akaike Information Criterion (NAIC) value. Performance of the model for hold-out-data, carried out on the basis of absolute percentage errors, is found to be extremely good. As far as forecasting out-of-sample data points is concerned, an optimal non-linear predictor was derived theoretically and used for forecasting India's marine products export data for the years 2002- 03 and 2003-04.

References

Draper, N. R. and H. Smith, 1998. *Applied Regression Analysis*, 3rd edn. John Wiley.

Ghosh, H., Prajneshu and G. Sunilkumar, 2006. Modelling and forecasting of bilinear time-series using frequency domain approach. *J. Combi. Info. Sys. Sci.*, 31: 121-37.

Gopakumar, K., S. Ayyappan, J. K. Jena, S. K. Sahoo, S. K. Sarkar, B. B. Satapathy, and P. K. Nayak, 1999. *National Freshwater Fisheries Development Plan*. Central Institute of Freshwater Aquaculture, Bhubaneswar.

Huang, Y. and N. R. Draper, 2003. Transformations, regression geometry and R^2. *Compu. Statist. Data Anal.*, 42:647–64.

Kandala, V. M. and Prajneshu, 2004a. Fuzzy von Bertalanffy growth model for determining age-length relationship. *Ind. J. Fish.*, 51:55-59.

Kandala, V. M. and Prajneshu, 2004b. Fitting allometric model using fuzzy least squares. *J. Mar. Biol. Assn.*, 46:120-23.

Kvalseth, T. O., 1985. Cautionary note about R^2. *Amer.Statistician*, 39:279-85.

McCausland, W. D., E. Mente, G. J. Pierce, and I. Theodossiou, 2006. A simulation model of sustainability of coastal communities: Fisheries, fishing, environment and labour markets. *Ecol. Modelling*, 193:271-94.

Prajneshu and K. P. Chandran, 2005. Computation of compound growth rates in agriculture: Revisited. *Ag. Econ. Res. Rev.*, 18:317-24

Prajneshu and R. Venugopalan, 1997. Statistical modelling for describing fish production in India. *Ind. J. Ani. Sci.*, 67:452-56.

Prajneshu and S. Ravichandran, 2001. Use of expected-value parameters for estimating length-weight relationship in fishes. *Ind. J. Fish.*, 48:199-202.

Prajneshu, 1991. Cautionary note about nonlinear models in fisheries. *Ind. J. Fish.*, 38:231-33.

Prajneshu, 2005. Statistical modelling in fisheries: A review. *Ind. J. Ani. Sci.*, 75:1008-12.

Ratkowsky, D. A., 1990. *Handbook of Nonlinear Regression Models*, Marcel Dekker.

Ravichandran, S. and Prajneshu, 2000. Modelling and forecasting of India's shrimp export. *J. Mar. Biol. Assn.*, 42:178-81.

Ravichandran, S. and Prajneshu, 2001. Structural time-series modelling for describing fish production. *Ind. J. Ani. Sci.*, 71:499-501.

Scott, A. and C. J. Wild, 1991. Transformations and R^2. *Amer. Statistician*, 45:127-28.

Seber, G. A. F. and C. J. Wild, 2003. *Nonlinear Regression*, 2nd edn. John Wiley.

Singh, R. K., Prajneshu and Ghosh, H., 2007. A two-stage fuzzy least squares procedure for fitting von Bertalanffy growth model. To appear in *Ind. J. Fish.*, 54.

Terdik, G., 1999. *Bilinear Stochastic Models and Related Problems of Nonlinear Time Series Analysis*. Springer-Verlag.

Venugopalan, R. and Prajneshu, 1997a. A generalized allometric model for determining length-weight relationship. *Biom. J.*, 39:733-39.

Venugopalan, R. and Prajneshu, 1997b. Von Bertalanffy growth model with autocorrelated errors. *Ind. J. Fish.*, 44:63-67.

Venugopalan, R. and Prajneshu, 1998. On nonlinear procedure for obtaining length-weight relationship. *Ind. J. Ani. Sci.*, 68:70-72.

Venugopalan, R. and Prajneshu, 2000. Pella-Tomlinson nonlinear statistical model with autocorrelated errors. *J. Ind. Soc. Ag. Statist.*, 53: 12-19.

Willmann, R., 2005. Economic aspects in the relationship between capture fisheries and fisheries. Paper presented in "Regional Workshop on Low Value and Trash Fish in the Asia–Pacific Region", Hanoi, Vietnam, 7-9 June.

Modelling, Forecasting, Artificial Neural Network and Expert System in Fisheries
Editors: Ajit Kumar Roy & Niranjan Sarangi
Published by: DAYA PUBLISHING HOUSE (2009)

Pages 13–26

Chapter 2

Is Fisheries Sector in India Changing its Gear?

B. Ganesh Kumar

Introduction

Fisheries, a sunrise sector in India, has recorded a faster growth than that of crop and livestock sectors (Kumar *et al.*, 2006). The sector contributes to the livelihood of a large section of economically underprivileged population of the country. It has been recognized as a powerful income and employment generator as it stimulates growth of a number of subsidiary industries and is a source of cheap and nutritious food besides being a foreign exchange earner (Ayyappan and Krishnan, 2004). With the changing consumption pattern, emerging market forces and technological developments, it has assumed added importance in India. It is undergoing rapid transformation and the policy support, production strategies, public investment in infrastructure, and research and extension for fisheries have significantly contributed to the increased fish production. Particularly, after the mid-1980s, the development of carp polyculture technology has completely transformed the traditional backyard activity into a booming commercial enterprise. Approximately 50 million people worldwide depend on fishing for all or most of their family earnings, while another 150 million depend on fish processing and the fleet servicing industry. More than 10 million work on 2.5 million small-scale fishing vehicles and account for 50 per cent of world's catch (FAO 2001).

India's Share in Fish Production

Fish production in the world rose from 23.50 million tonnes in 1950-51 to 140.48 million tonnes in 2003-04. Correspondingly, the fish production in India has touched 6.40 million tonnes in 2003-04 from a mere 0.75 million tonnes in 1950-1951

(Table 2.1). The total fish production of our country stood at 6.57 million tonnes in 2005-06. The share of India in global fish production has grown gradually from about 2.66 per cent during the 1960s and 1970s to 4.56 per cent in 2003-04.

Table 2.1: Fish Production in World and India, 1950-51 to 2003-04
(*Million tonnes*)

Year	*World*	*% Change*	*India*	*% Change*	*India's Share*
1950–51	23.50	–	0.75	–	3.19
1960–61	43.60	85.53	1.16	54.67	2.66
1970–71	66.20	51.83	1.76	51.72	2.66
1980–81	72.30	9.21	2.44	38.64	3.37
1990–91	98.26	35.91	3.84	57.38	3.91
2000–01	129.00	32.35	5.66	47.40	4.39
2003–04	140.48	8.90	6.40	13.07	4.56

Source: ICAR-ICLARM Project on 'Strategies and options for increasing and sustaining Fisheries and Aquaculture production to benefit poor households in India, 2004 and Economic survey 2006-07, GOI.

It shows that growth in fish production in India has been at a faster rate than in world; mainly due to increasing contribution from inland fisheries. It can also be seen in Figure 2.1 that in the pre-WTO period of 1990-91, the share of India's fisheries was 3.90, which rose to 4.56 in the post-WTO period of 2003-04. Overall, the share of developing world in the total world fish production increased from 43 per cent in 1973 to about 73 per cent in 1997 (Figure 2.2), which has been mainly due to the increasing contribution from countries like China and India (Delgado *et al.*, 2003).

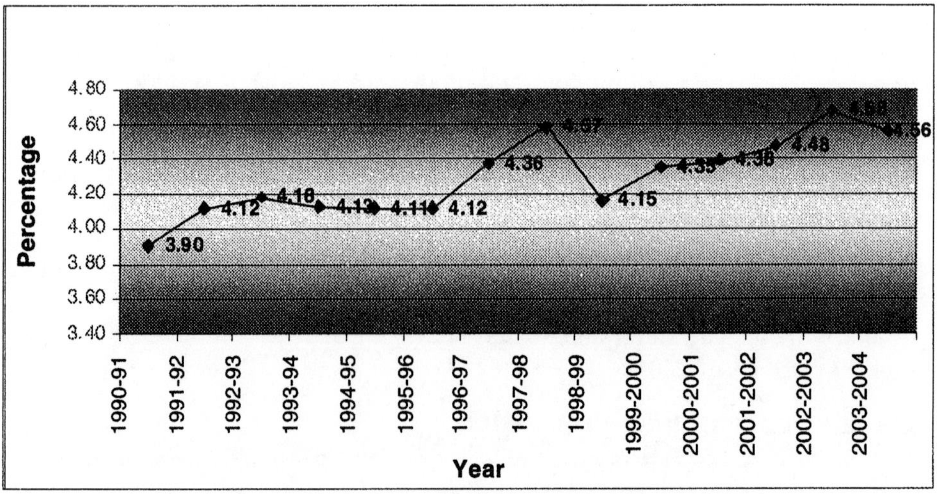

Figure 2.1: India's Share in World Fish Production

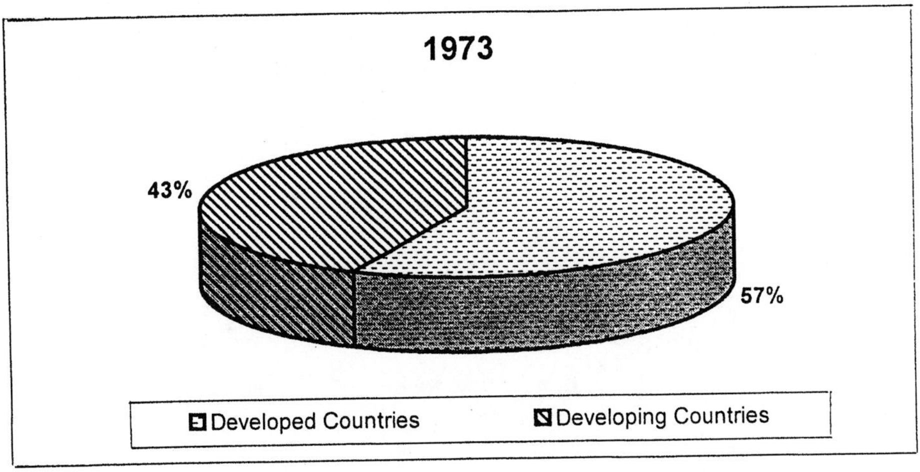

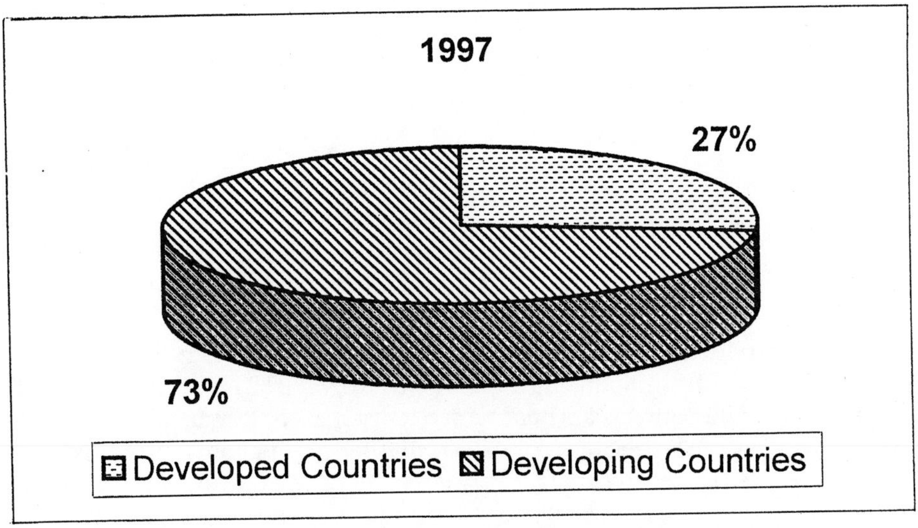

Figure 2.2: Changing Share in Fish Production of Developing and Developed Countries for Food in 1973 and 1997
***Source*: Delgado *et al.* (2003)**

Contribution to Indian Economy and Prospects of Fisheries Growth

With fisheries sector comprising marine fisheries, freshwater and brackish water aquaculture and inland fisheries consisting of tanks and reservoirs, the potentiality of this sector as a whole remains to be fully tapped and it remains a sector of much promise. The fisheries sector in particular is more complex enterprise that functions under integrated network of natural resources, other enterprises with forward and backward linkages with fisheries and other socio-political variables. The major functions of fisheries enterprises, *viz.* production, transportation, storage and processing involve value addition from labour, capital and management, which significantly influence the rapid economic development of the country (Ayyappan and Krishnan, 2004). In the last 25 years, unlike agriculture, the contribution of fisheries sector to Gross GDP continued to grow at a rapid pace because of expansion of culture fisheries enterprise. The share of agriculture and allied activities in the total GDP is constantly declining. It was 34.69 per cent in 1980-81 and declined gradually to become 17.62 per cent in 2004-05. In contrast, the contribution of fisheries sector to the total GDP has gone up from 0.75 per cent in 1980-81 to 1.04 in 2004-05 (at current prices).

Similarly, the share of fisheries in agriculture GDP (AgGDP) has increased robustly from 2.17 per cent in 1980-81 to 5.93 per cent in 2004-05. This sector is in fact pushing the agricultural growth upward for the past 5 and half decades.

Table 2.2: Comparison and Growth of Fisheries Sector

Period	*Percent Contribution of*		
	Agriculture to Total GDP	*Fish to Total GDP*	*Fish to Ag GDP*
1980–81	34.69	0.75	2.17
1990–91	28.42	0.96	3.37
2000–01	22.26	1.18	5.32
2004–05	17.62	1.04	5.93

Source: National Account Statistics, CSO, GOI.

Fish Production: Structure and Trend

The fish production in India witnessed a spectacular growth since independence. It rose from a mere 0.75 million tonnes in 1950-51 to over 6.57 million tonnes in 2005-06 (Table 2.3). In the initial years, marine sector used to contribute more to total fish production than inland sector. In 1950-51, marine production contributed about 71.01 per cent, which fell gradually to 42.77 per cent in 2005-06, while inland sector started contributing from 28.99 in 1950-51 to about 57.23 per cent in 2005-06. In fact, by the year 2000, its share crossed 50 per cent and continue to improve further in the coming years (Figure 2.3).

Table 2.3: Fish Production by Source in India

Year	*Fish Production (Mt)*		
	Marine	*Inland*	*Total*
1950–51	0.534	0.218	0.752
1960–61	0.880	0.280	1.160
1970–71	1.086	0.670	1.756
1980–81	1.555	0.887	2.442
1990–91	2.300	1.536	3.836
2000–01	2.811	2.845	5.656
2003–04	2.810	3.760	6.570

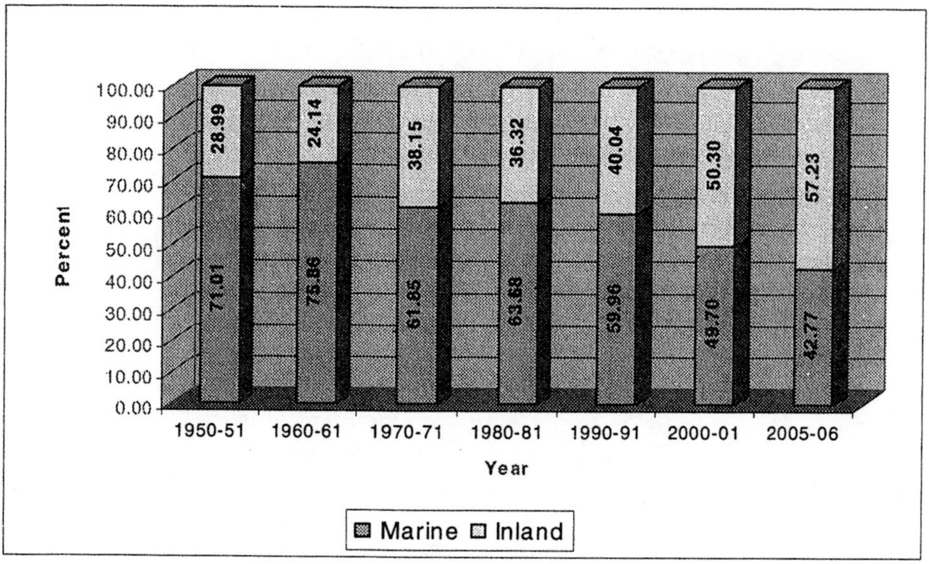

Figure 2.3: Fish Production by Source in India in per cent

Expansion of fleet capacity, technological innovation, and increases in investment led to explosive growth in the exploitation of marine fisheries through the 1960s, 1970s and 1980s. But from the late 90s onwards, the marine fisheries production has reached a plateau and it seems that it can register only marginal increase in the near future. With most wild fisheries near maximum sustainable exploitation levels, capture fisheries will most likely to grow slowly.

Table 2.4: Compound Growth Rate in Fish Production, 1980-81 to 2005-06

Year	*Marine*	*Inland*	*Total*
1980–81 to 1989–90	3.80	5.28	4.39
1990–91 to 1999–00	2.33	6.55	4.13
2000–01 to 2005–06	–0.21	5.37	2.75
1980–81 to 2005–06	3.24	6.20	4.60
1980–81 to 1980–91 (Pre–WTO)	4.35	5.43	4.78
1991–92 to 2005–06 (Post–WTO)	0.84	5.71	3.18

On the other hand, inland fish production was on constant rise. The inland fisheries include both capture and culture fisheries. The capture fisheries have been the major sources of inland fish production till mid-1980s. But the fish production from natural waters like rivers, lakes, etc. followed a decling trend, primarily, due to proliferation of water control habitat degradation (Katiha and Bhatta, 2002). The depleting resources, energy crisis and resultant high cost of fishing etc. have led to increased realization of the potential and versatility of aquaculture on a sustainable and cost-effective alternative to capture fisheries.

In the last 25 years, total fish production has been growing at an annual growth rate of about 4.60 per cent, in which marine sector was growing at a rate of 3.24 and inland sector was growing at a rate of 6.20 per cent.

In all, inland sector fared better in all the periods, *viz.* 1980-81 to 1989-90, 1990-91 to 1999-00 and 2000-01 to 2005-06. There seem to be a slower pace in growth of this sector in the recent times. In contrast, marine sector is witnessing a negative growth rate during the period from 2000-01 to 2005-06, which indicates the exhaustion of marine resources especially in the in shore and near shore water, where maximum harvest has happened. About 90 per cent of the present production from the marine sector is within a depth range of upto 50-70m and the remaining 10 per cent from depth extending upto 200m. While 93 per cent of the production is contributed by artisanals and motorized sectors, the remaining 7 per cent is contributed by deep sea fishing fleets confining their operation mainly to the shrimp grounds in the upper east coast. Hence, in order to enhance and sustain the contribution from this sector, we need to go deep sea for targeting its untapped potential that is supported by enhanced investment in mechanized vessels, capacity strengthening of artisanals and probably a proper institutional structure to share the benefits.

The growth rates in pre and post WTO periods were also estimated. It is noticed that the pre-WTO period witnessed an impressive growth rate of about 4.78 as compared to post WTO period (3.18). This trend was mainly due to the marine sector, which is understandable by the fact that the country's fish export basket was dominated by marine species and buoyance of marine export might have propelled the growth of marine catch, and vice versa. The post-WTO period imposed many quality regulations in terms of SPS measures on developing countries like India, which couldn't create huge investment in the infrastructures required to produce export-quality marine fisheries products that are acceptable to our trading partners, especially EU, USA and Japan. In contrast to the marine sector, inland sector continued to grow better in the post-WTO period also, which is possibly because of enhanced public and private investment for inland fisheries sector especially through different development programmes and research by the Government of India since IV plan onwards that started delivering results continuously. Figure 2.4 clearly depicts the dominance of inland fisheries sector in the contribution to total fish of the country, particularly from 1990s.

To assess the trend of fish production in the country, the last 25 years data (from 1980-81 to 2005-06) was used. Both linear and non-linear trend analysis was tried and the best equation fitted. The total fish production in India has been following a linear trend (Figure 2.5) and it is likely proceed further in the same direction. The trend pattern of marine and inland fish production revealed that while the marine sector's production was increasing at a decreasing rate, the inland sector's production was increasing at an increasing rate, the possible reasons for which are that inland aquaculture activities are gaining much importance in some of the states like Uttar Pradesh, Andhra Pradesh, Punjab etc. in recent years.

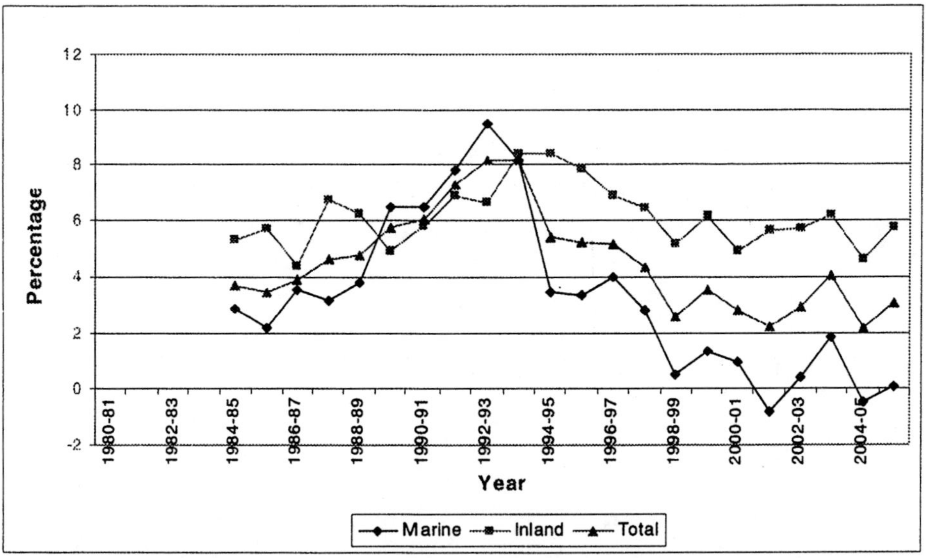

Figure 2.4: Annual Growth Rate (Per cent) in Fish Production, 1980-81 to 2005-06 (5 Yearly Averages)

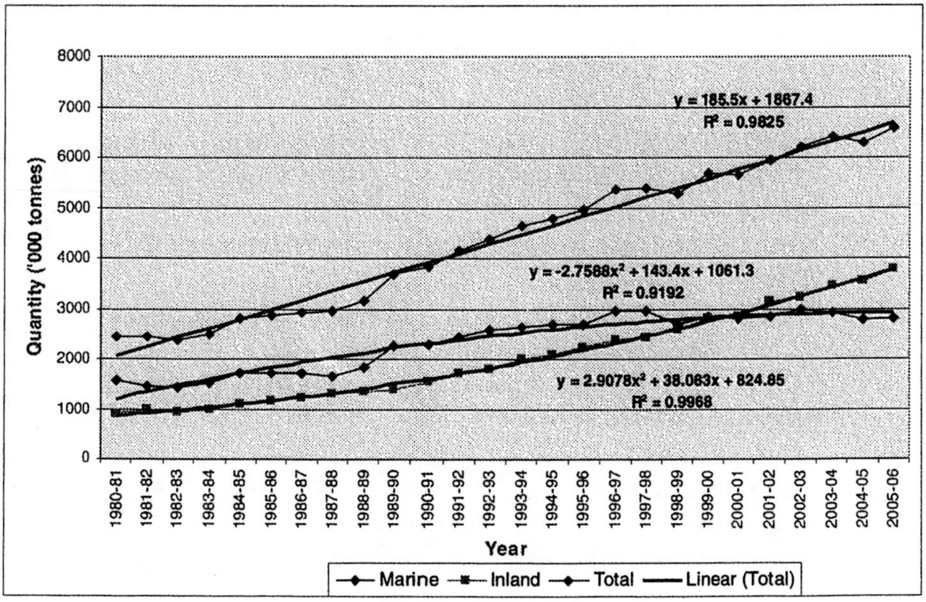

Figure 2.5: Trend in Fish Production, 1980-81 to 2005-06

The trend equations for both marine and inland fish production, as well as total fish production are given below:

Marine: $Y = -2.7588X^2 + 143.4X + 1061.3$

Inland: $Y = 2.9078X^2 + 38.08X + 824.85$

Total: $Y = 185.5X + 1867.4$

Table 2.5: Fish Production in 2020

Year	*Fish Production (Mt)*		
	Marine	*Inland*	*Total*
2010	2.88 (38.61)	4.58 (61.37)	7.46
2015	2.70 (32.07)	5.72 (67.33)	8.42
2020	2.38 (25.37)	7.00 (74.63)	9.38

Note: Parentheses indicates percentage to total.

Based on the trend observed in the last 25 years, assuming that the same scenario would prevail on both public and private investment pattern in the sector, the fish production of the country would reach 7.46, 8.42 and 9.38 million tonnes in 2010, 2015 and 2020 respectively.

In the coming decades, aquaculture will likely be the greatest source of increased fish production, as fish farmers expand the water surface area under cultivation and increase yields per unit of area cultivated.

But the sector must overcome several major challenges, if it has to sustain the rapid growth of the past 25 years.

- ✩ Aquaculture has to face competition from others users of land and water, as these resources would become more scarce in future (Rosegrant *et al.*, 2002)
- ✩ Marine sector would face energy crisis as it would require more fossil fuels for further expansion.
- ✩ The aquaculture production would be restrained by diseases as the sector expands (Subasinghe, *et al.*, 2001).
- ✩ The availability of fish meat and fish oil from wild caught feed inputs may also become a limiting factor.

Yield increases can come either from increased inputs or greater efficiency of inputs. It is likely that in the next several decades, aquaculture production will benefit from both these sources of yield growth. Greater use of compounded aquaculture feeds along with improvements in rearing technology and selective breeding has the potential to significantly increase the productivity of many farm of aquaculture.

Fish Production in India from I Plan to X Plan

The total fish production in our country was hovering around 1 million ton since the end of third five year plan. Until then, the policy of the country was focused on enhancing fish production only, with little attention on such issues as marketing, storage, transportation, etc. Later, emphasis was also laid on creating facilities for ice-cold storages, processing and canning. As a result of this, the fish production started increasing significantly from VI Five Year Plan onwards during which it attained the total increase of about 47.11 per cent. Its impressive growth continues all through subsequent plan periods also. Overall, plan-wise fish production is witnessing an exponential growth path.

The trend equation for plan-wise fish production is as follows

$$Y = 567.34e^{0.2332X}$$

This trend confirms the characteristic of this sunrise sector in Indian agriculture in days to come.

In fact, the X Plan proposed a fish production target of 8.19 Mt (Table 2.6), envisaging a growth rate of 5.44 per cent per annum (marine 2.5 per cent and inland 8.0 per cent). During the X Plan, new initiatives for development of fisheries were planned to increase production and productivity from deep sea, inland capture fishery resources like rivers, canals, etc., culture sources like reservoirs, beels, ox-bow lakes, etc. and measures for replacement of fishery resources through mariculture. Besides development of infrastructural facilities or a better post-harvest management, technology for sustainable aquaculture, setting up of cold storage and marketing network through viable fisheries co-operatives, etc. were to be taken up to ensure better livelihood for fishers and enhance export promotion for economic development of the country (X Five Year Plan, 2002-07 documents, Planning Commission).

As against the targeted production of 8.19 Mt, the country produced about 6.57 Mt until 2005-06 with inland sector contributing about 57.23 per cent and marine

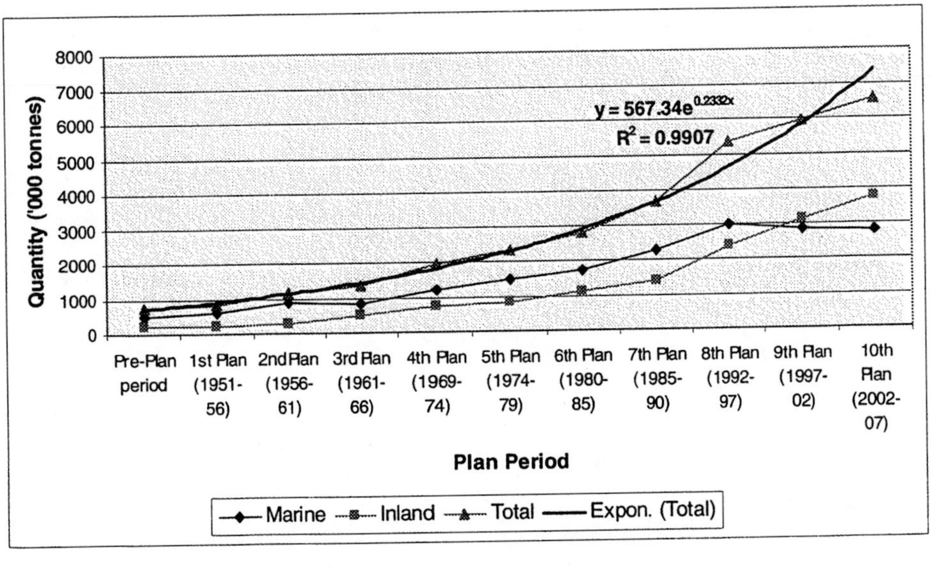

Figure 2.6: Trend in Fish Production (Plan-wise)

sector 42.77 per cent. Both sectors witnessed an annual growth rate of 5.1 per cent and–2.4 per cent, respectively as against 8 per cent and 2.5 per cent set by the Planning Commission.

Table 2.6: Targets and Achievements in Production and Growth Rate in X Plan

X Plan	*Marine*	*Inland*	*Total*
Production (Mt)			
Target	–	–	8.19
Achievement	–	–	6.57
Growth Rate (per cent)			
Target	2.5	8.0	5.5
Achievement	–2.4	5.1	1.6

Growth of Fisheries Sector in Different States of India

The growth rate analysis for various states from 1990-91 to 2003-04 showed that fish production had a significant growth in all states except Karnataka and Madhya Pradesh. In some of the states like Andhra Pradesh, Karnataka, Kerala, Maharashtra, Orissa and West Bengal the growth of inland fisheries was found to be higher than that of marine fisheries. But in Gujarat and Tamil Nadu, marine fisheries growth was observed to be more than that of inland fisheries.

Table 2.7: Growth Rate of Fish Production by States, 1990-91 to 2003-04

State	*Marine*	*Inland*	*Total*
Andhra Pradesh	5.57**	13.41**	10.38**
Assam	–	3.53**	3.53**
Bihar	–	3.51**	3.51**
Gujarat	1.67*	–0.80 NS	1.51*
Haryana	–	4.49**	4.49**
Himachal Pradesh	–	1.84**	1.84**
Jammu and Kashmir	–	3.18**	3.18**
Karnataka	–0.84 NS	4.32*	0.87 NS
Kerala	1.33**	6.90**	1.83**
Madhya Pradesh	–	1.21 NS	1.21 NS
Maharashtra	1.18 NS	6.06**	2.13**
Orissa	2.04 NS	5.34**	3.83**
Punjab	–	15.31**	15.31**
Rajasthan	–	5.73**	5.73**
Tamil Nadu	1.99**	1.52*	1.88**
Uttar Pradesh	–	7.29**	7.29**
West Bengal	2.69**	4.47**	4.16**
Delhi	–	1.98 NS	1.98 NS

**: Significant at 1 per cent; *: Significant 5 per cent; NS: Non-significant.

Share of States in Total Fish Production

Marine Fish Production

The share of various states in the country's total marine production was analysed and its composition was presented for the year 1990-91 and 2003-04 to understand the changing share of various states in this crucial export oriented sector (Figure 2.7).

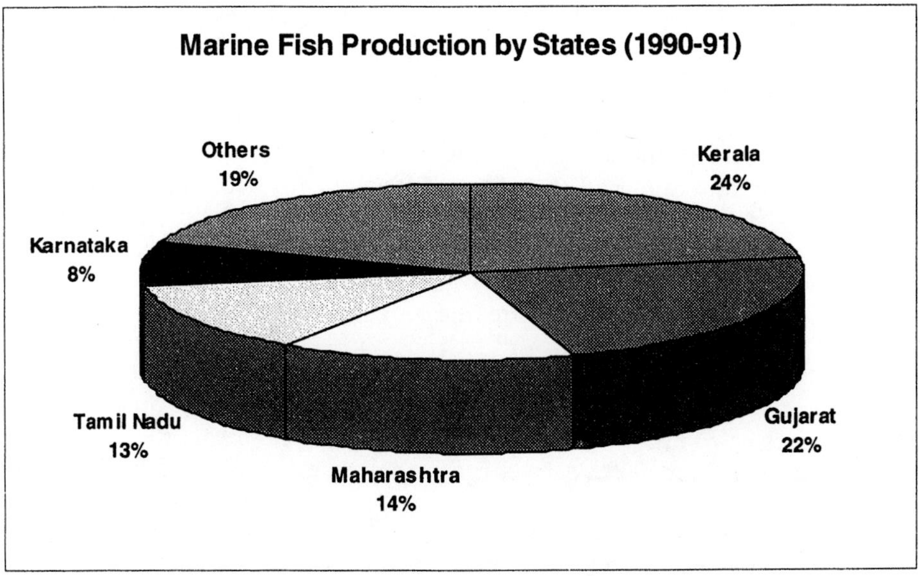

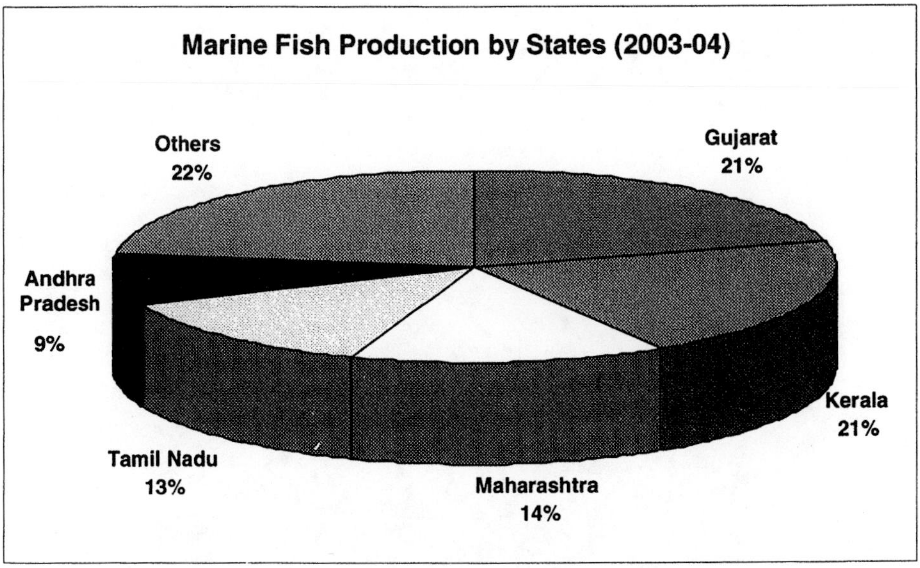

Figure 2.7: Changing Share of States in Marine Production, 1990-91 and 2003-04

In 1990-91, Kerala, Gujarat, Maharashtra, Tamil Nadu and Karnataka together contributed about 81 per cent in the total marine production. In the last one and half decade, Gujarat was reaching the plateau and its contribution was reducing recently and Kerala is static in its increase, while the Maharashtra and Tamil Nadu's share didn't change. But, the one state that has been increasing at a rapid pace in marine production is Andhra Pradesh, which replaced Karnataka in its share to total production in the last 15 years. Among the others states, which are contributing significantly to total marine production, are West Bengal and Orissa and among UTs, the major contributors are Pondicherry, and Andaman Nicobar islands, though their current production is stagnating in the recent years.

It is reasonably known that marine fisheries are probably near the ceiling of the potential. However, scientific opportunities may exist in area like mariculture of filter feeders. But, the development of this sector is beset with major management problems. It is facing the problem of influx of sewage water and other pollutants causing health hazards. Yet another problem faced relates to global warming. The marine fisheries also have to bear large-scale destruction of juvenile fishes. Hence, institutionalization of conservatory and regulatory/control measures are important.

Inland Fish Production

The share of various states in the country's total inland fish production was analyzed and its composition was presented for the year 1990-91 and 2003-04 to understand the dynamics of this sector in contributing to total fish production (Figure 2.8).

Figure 2.8: Changing Share of States in Inland Production, 1990-91 and 2003-04

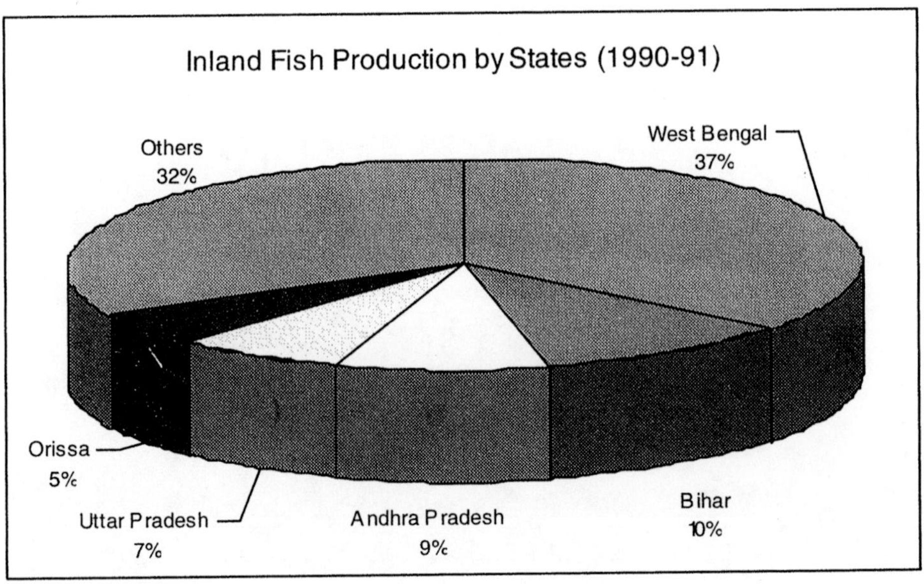

Contd...

Figure 2.8–Contd...

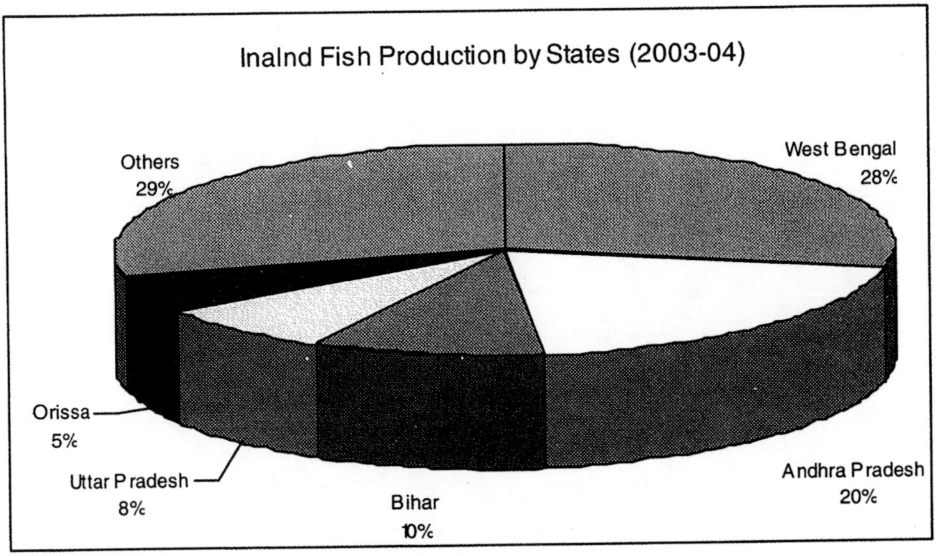

In the year 1990-91, the major producing states were West Bengal, Bihar, Andhra Pradesh, Uttar Pradesh and Orissa. They together contributed about 68 per cent to total inland fish production. In the subsequent one and half decades, Andhra Pradesh had made tremendous progress in its share to total inland production than West Bengal, which was static in its increase and hence lost to the emerging states. Bihar and Orissa were maintaining their share in inland production, though they have reached stagnation in its output. Another major state, whose contribution is improving rapidly, is Punjab. Apart from these, the states which are reaching a point of maximum output are Assam, Kerala, Maharashtra and Tamil Nadu. Some states were also experiencing negative growth in its production and they are Gujarat, Karnataka and Madhya Pradesh. Again the promising UT in inland production is Pondicherry.

References

Ayyappan, S. and Krishnan, M. (2004). Fisheries sector in India: Dimensions of development. *Indian Journal of Agricultural Economics*, 59 (3): 392-412.

Delgado, C.L., Wada, N., Rosegrant, M.W., Meijer, S. and Ahmed, M. (2003). The future of fish: Issues and trends to 2020. Book published by IFPRI and World Fish Centre. ISBN: 0-89629-725-X.

Katiha, P. K. and Bhatta, R. C. (2002). Production and consumption of aquacultural products in India: Past trend, present status and future prospects. Presented at Special session on Strategies and Options for Sustainable Aquacultural Development at World Aquaculture at 2002, Beijing, China, 23-27 April, 2002.

Kumar, P., Dey, M. M. and Paragnas (2006). Fish supply projections by production environments and species types in India. *Agricultural Economics Research Review*, 19 (2): 327-351.

Rosegrant, M., Cai, X. and Cline, S. (2002). World water and food to 2025: Dealing with scarcity. Washington, D.C.: IFPRI.

Subasinghe, R. P., Bondad-Reantaso, M. G. and McGladdry, S. E. (2001). Aquaculture development, health and wealth. *In* : Technical Proceedings of the Conference on Aquaculture in the Third Millennium, Bangkok, Thailand, 20-25, February, 2000.

Modelling, Forecasting, Artificial Neural Network and Expert System in Fisheries
Editors: Ajit Kumar Roy & Niranjan Sarangi
Published by: DAYA PUBLISHING HOUSE (2009)

Pages 27–36

Chapter 3

Non-Linear Models for Forecasting Fish Production from Ponds

Lalmohan Bhar

Introduction

Reliable and timely forecasts of fish production will be of immense use to planners and policy makers. Such advance information is useful in formulating import and export policies and exercising several administrative measures for storage and marketing of fish production. In the present chapter some aspects of development of forecasting techniques for fish production will be discussed. In general, commercially fishes are reared for a year or so. It will be very much helpful for the farmers if they know the average fish weight at harvest at least two to three months back. For forecasting purpose some data on fish weights taken daily/weekly/monthly are required. Since, these data are collected over time, no linear model would be appropriate. The appropriate models should be non-linear. Walia *et al.*(1997) built logistic model by taking the monthly fish weight as response variable. Though they considered the basic model as non–linear, yet the intra–individual response variance was not considered accordingly. Here by individual, we mean pond. Observations are taken on fish weight from ponds. They assumed homogeneous error structure for the model. However, this assumption seldom calls upon in practice. In the present chapter we postulated appropriate model giving full emphasis on the structure of error variance.

Data Description

The forecasting techniques, using some real data have been demonstrated. The data have been collected from an experiment to study the development, growth and reproductive biology of culture fishes in tanks. These data are part of the above study conducted by researchers of Fisheries Research Complex, Punjab Agricultural University, Ludhiana, India. Data are collected on three species of fish, *viz.* Rohu, Mrigal and Common Carp. Monthly weights are recorded from October, 1982 to July, 1984. Thus there are in total 22 data points for each species. For recording of fish weight random samples of five specimens of each species from each tank were taken at monthly intervals and their weights were measured and average weight is taken as the monthly fish weight. For demonstration we use data on Rohu only. Cemented tanks each with an area of 0.009 ha (13.6mx5.4mx1.22m) were selected for the experiment. Before filling them with water, a thin layer of solid was spread at the bottom of each tank. Tanks were filled with tube–well water, the level of which was maintained almost at 1 m.

Stocking of Rohu was done in October, 1982, @7927 per ha. Average weight at the time of stocking was 1.0 gram. In the present demonstration, we have taken fish weight data on Rohu collected from controlled tanks in which no fertilizer was added and from some fertilized tanks in which Farm Yard Manure (FYM) and Poultry Manure (PM) @ 20,000 kg/ha/yr and in combination of 1:3 are used. Final weight was recorded on 7th July, 1984 at the time of harvest.

Full data points (22 in number) have been used to develop models for growth analysis. For developing forecasting models similar kind of models have been developed using partial data. The models are fitted with 17, 18 and 19 data points respectively. Remaining data points are kept for validation purpose. First we discuss the modelling and forecasting of fish weight data on Rohu collected from controlled tanks, taking into account the heterogeneity of error variance in the model. However, sometimes assumption of heterogeneous error structure may not be a serious restriction for forecasting. Assumption of homogeneous error structure may well serve the purpose. We, therefore, demonstrate this modelling technique using data collected from fertilized tanks.

Objectives and Model Formulation

References on modelling with repeated measurement data is vast in the literature. A good account of the modelling approach of such kind is found in Zeger et al (1988), Beal and Sheiner (1988), Davidian and Giltinan (1993), Gumpertz and Pantula (1992), Jenrich and Sampson (1986), and van Houwelingen (1988). Examples of some applications are found in Aziz *et al.* (1978), Gregoire and Schabenberger (1994), Walters *et al.* (1985) and Brillinger (1987). Though a vast literature on modelling with repeated measurement data is available in the literature, yet not much work seems to be available on forecasting using such data.

Inspection of the scatter plots of fish weight data on Rohu reveals that the usual logistic function:

$$y = \frac{\beta_1}{1 + \beta_2 EXP(\beta_3 x)},\ \beta_1,\ \beta_2 > 0,\ \ \beta_3 < 0 \qquad (1)$$

Provides a reasonable representation of average fish weight y and time x. It is evident from the scatter plots that considerable variation in parameters β_1, β_2 and β_3

that characterize the growth pattern exists among the species. For this kind of growth data, it is reasonable to expect serial correlation among measurements within the same species. In addition, intra–species variability may be expected to increase with the average level of response. More specifically, let y_j denote the response taken at the jth covariate value x_j, $j = 1, 2, \ldots, n$. In the present case, the response vector $\mathbf{y} = (y_1, \ldots, y_n)$ summarizes the information available for a single species taken at values $\mathbf{x} = (x_1, \ldots, x_n)$ of time variable. Thus the model for the jth observation can be written as

$$y_j = f(x_j, \beta), \tag{2}$$

where, $f(x_j, \beta) = \dfrac{\beta_1}{1 + \beta_2 EXP(\beta_3 x_j)}$ and $\beta = (\beta_1 \quad \beta_{21} \quad \beta_3)'$.

The random errors e_j reflect uncertainty in the measured response. Thus $\mathbf{e} = (e_1 \quad e_2 \quad \ldots \quad e_n)'$ summarizes the uncertainty for all observations on given species.

The classical nonlinear regression frame work specifies that data arise according to (2) along with some assumptions such as the errors are uncorrelated, having mean zero and common variance σ^2.

Assumption of constant intra–individual response variance is violated frequently in practice, particularly for growth data of present kind, instead they exhibit constant coefficient of variation rather than constant variance; that is, variance is proportional to the square of the mean response. Thus a more appropriate assumption would be

$$E(y_j) = f(x_j, \beta), \ \text{var}(y_j) = \sigma^2 \{f(x_j, \beta)\}^2 = g_j \ (\text{say}), \tag{3}$$

where, the scale parameters is the coefficient of variation.

The parameters of the logistic growth model have interpretations that pertain directly to the study objectives. The function (1) may be derived by assuming that rate of growth relative to current size is in constant proportion, by a factor to remaining growth to be attained (*e.g.* Draper and Smith, 1981). Consequently, β_3 plays the role of a growth rate constant so that comparison of growth rate among fish species may be based on this parameter. β_1 represents the limiting growth value.

As discussed earlier, for this type of growth data, intra–species variance that increases systematically with average fish weight is likely. Studentized residuals from separate ordinary least squares fits of the logistic model are plotted against predicted values for the species. These plots also confirm this assumption.

Correlation among observations on a given tank is especially likely in the present context. To accommodate intra–pond correlation, a description of the assumed correlation pattern among the elements of e_j' s; when the observations are taken over time, standard models for serial correlation patterns are available, for example, the auto–regressive (AR) model of order one. In the present case, the n measurements are equally spaced in time, we, therefore, assume that the correlation between two adjacent observations is constant.

Based on the above considerations, we fit the above mentioned logistic model to the fish data using appropriate approach that allow straightforward accommodation of these feature: Generalized least squares (GLS) estimation of individual regression parameters incorporating pooled estimation of within tank covariance parameters. Estimation procedure was implemented in the SAS matrix programming language (IML).

Methods

Since the heterogeneity of variance is evident, ordinary least square estimation of β may be inefficient. However, classical least squares method could be modified to incorporate the heterogeneity into model as follows:

In our model, coefficient of variance is constant, thus variance is known up to the value of the regression parameter p, which appears through the mean response. An obvious approach is thus to take advantage of the functional form for variance to construct estimated weights, replacing β by a suitable estimate, and to apply the weighted least squares idea. Variance of y_j are known upto a constant of proportionality; that is var $(y_j) = \frac{\sigma}{w_j}$, where $w_j = \frac{1}{f(x_j, \beta)}$. Now, with assumption of independent, normal errors, it is known that the maximum likelihood estimation of β is the value $\hat{\beta}_{WLS}$ that minimizes

$$\sum w_j \{y_j - f(x_j, \beta)\}^2 \tag{4}$$

This method is known as weighted least squares (WLS) estimation and estimator for σ^2 is usually replaced by $\hat{\sigma}^2_{WLS} = \frac{1}{n-p}\sum w_j \{y_j - f(x_j, \hat{\beta}_{WLS})\}^2$. Ordinary Least Squares is a natural choice to use for construction of estimated weights. Thus the estimated weights are formed as $\hat{w}_j = \frac{1}{f^2(x_j, \hat{\beta}_{OLS})}$, then we solve for (4), using the $\hat{w}_j$ in place w_j. This is actually a special case of generalized least squares (GLS) methods:

(*i*) Estimate β by a preliminary estimator $\hat{\beta}^{(p)}$, *e.g.*, the OLS estimator $\hat{\beta}_{OLS}$.

(*ii*) Form estimated weights $\hat{w}_j = \frac{1}{f^2(x_j, \hat{\beta}_{OLS})}$

(*iii*) Using the weights from step (*ii*), re-estimate β by weighted least squares. Treating the resulting estimator as a new preliminary estimator, return to step (*ii*). Continue this process till convergence is met.

Step (*iii*) provides for the possibility of iteration.

Once the weights are formed through above method, we estimate the covariance parameter through AR(I) method. Then general variance–covariance matrix (W) of y becomes known. We now apply GLS to obtain the final estimates of β denoted by $\hat{\beta}_{GLS}$.

Newton–Raphson algorithm has been used to compute the estimates. For developing the models under the assumption of homogeneous error structure, above method is skipped and SAS-IML is used to estimate the parameters.

Goodness of Fit Statistics

Using following statistics comparison among the models has been done. The most popular statistic used is R^2.

$$R^2 = 1 - (\text{Residual sum of squares}/\text{Total sum of squares}),$$

where, Residual sum of squares is $\sum_{i=1}^{n}(Y_i - \hat{Y}_i)^2$, Total sum of squares is $\sum_{i=1}^{n}(Y_i - \overline{Y})^2$,

Y_i dependent variables (weights), $\hat{Y}_i$ is predicted values or estimated value and $\overline{Y}$ is mean of dependent variables.

Also, it was suggested that the nonlinear model should not be judged as the best model only on the basis of R^2, and one should go for the measures like RMSE and MSE for judging the goodness of fit of the model.

$$\text{Root mean squared error (RMSE)} = \left[\sum_{i=1}^{n}\frac{\left(Y_i - \hat{Y}_i\right)^2}{n-p}\right]^{1/2}$$

$$\text{Mean Square Error (MSE)} = \frac{\sum_{i=1}^{n}(Y_i - \hat{Y}_i)^2}{n-p},$$

where, n = number of observations and p = number of parameters in the model.

Results and Discussion

Table 3.1 summarizes the results for fitting of fish data on Rohu under the assumption of heterogeneity. Serial correlation coefficient estimated though AR(I) is 0.86. Positive high 1st order correlation coefficient is expected as fish weight increases with the time. An important point to note that there is a tremendous improvement in fitting under the assumptions of heteroscedasticity and heteroscedasticity with auto–correlation. Naturally we expect better result for forecasting when we assume heteroscedasticity or heteroscedasticity with auto–correlation. This is discussed in the next section.

Table 3.1: Fits of Fish Data (Full Data)

Species		*Model 1*		*Model 2*		*Model 3*	
		$\hat{\beta}_i$	*SE*	$\hat{\beta}_i$	*SE*	$\hat{\beta}_i$	*SE*
Rohu	$\hat{\beta}_1$	1541.68	80.19	2103.71	150.41	1781.04	79.24
	$\hat{\beta}_2$	140.42	30.45	58.33	7.48	82.04	9.69
	$\hat{\beta}_3$	0.30	0.01	0.21	0.01	0.24	0.01
	R^2	0.97		0.98		0.99	
	MSE	689.87		276.13		97.54	
	RMSE	18.95		13.74		8.06	

Model 1: Homoscedastic model.

Model 2: Heteroscedastic model.

Model 3: Heteroscedastic with Auto-correlation.

SE: Stands for standard error.

Forecasting Models

For forecasting purpose we develop separately three models using partial data. Tables 3.2–3.4 contain the results of these fitting using first 17, 18 and 19 data points respectively. Once again we see that there is an improvement in fitting of heteroscedastic with auto-correlation models over the fittings of homoscedastic models.

Table 3.2: Fits of Fish Data (Upto 19th Month) to Model 1, Model 2 and Model 3

Species		*Model 1*		*Model 2*		*Model 3*	
		$\hat{\beta}_i$	*SE*	$\hat{\beta}_i$	*SE*	$\hat{\beta}_i$	*SE*
Rohu	$\hat{\beta}_1$	943.13	32.02	1097.01	.66	1221.81	0.90
	$\hat{\beta}_2$	400.59	79.28	164.54	.49	97.65	0.32
	$\hat{\beta}_3$	0.45	0.02	0.36	*	0.36	*
	R^2	0.96		0.97		0.99	
	MSE	745.98		345.13		112.54	
	MAE	20.95		16.51		9.04	

Table 3.3: Fits of Fish Data (Upto 18th Month) to Model 1, Model 2 and Model 3

Species		*Model 1*		*Model 2*		*Model 3*	
		$\hat{\beta}_i$	*SE*	$\hat{\beta}_i$	*SE*	$\hat{\beta}_i$	*SE*
Rohu	$\hat{\beta}_1$	1030.39	36.29	1210.43	.63	1350.71	.89
	$\hat{\beta}_2$	319.37	65.68	128.09	.31	78.31	.21
	$\hat{\beta}_3$	0.42	0.20	0.33	*	0.28	*
	R^2	0.95		0.97		0.98	
	MSE	989.43		545.68		198.87	
	MAE	30.67		24.54		15.82	

Table 3.4: Fits of Fish Data (Upto 17th Month) to Model 1, Model 2 and Model 3

Species		*Model 1*		*Model 2*		*Model 3*	
		$\hat{\beta}_i$	*SE*	$\hat{\beta}_i$	*SE*	$\hat{\beta}_i$	*SE*
Rohu	$\hat{\beta}_1$	1128.51	42.39	1352.82	0.68	1560.28	1.14
	$\hat{\beta}_2$	256.16	54.54	99.70	0.20	60.53	0.14
	$\hat{\beta}_3$	0.39	0.02	0.29	*	0.24	*
	R^2	0.94		0.97		0.98	
	MSE	937.43		456.98		154.98	
	MAE	23.65		17.09		10.32	

Note: * Stands for negligible.

Validation of the Models

Validation of the developed forecast models was carried out using the remaining data. These results are presented in the Tables 3.5–3.7 respectively. Since we want to compare the results with the models developed under homoscedastic model we use the forecast error measure as

$$PE = \frac{\text{Observed} - \text{Forecast}}{\text{Observed}} X100$$

From the Tables it is evident that forecast models developed under the assumption of heteroscedastic error structure is better than that of models developed under the assumption of homoscedastic error structure and it is best under the assumption of heteroscedasticity with auto-correlation. Moreover, accuracy of forecast increases if the forecast period is reduced.

Table 3.5: Forecast of Fish Weights Using 19 Data Points

Species	*Month*	*Observed*	*Model 1*		*Model 2*		*Model 3*	
			Forecast	*PE*	*Forecast*	*PE*	*Forecast*	*PE*
Rohu	20	1120.3	1021.40	8.82	1074.35	4.10	1096.37	2.13
	21	1240.3	1053.70	15.04	1134.46	8.53	1173.01	5.42
	22	1360.3	1076.76	20.84	1183.64	12.98	1240.69	8.79

Table 3.6: Forecast of Fish Weights Using 18 Data Points

Species	*Month*	*Observed*	*Model 1*		*Model 2*		*Model 3*	
			Forecast	*PE*	*Forecast*	*PE*	*Forecast*	*PE*
Rohu	19	1010.3	931.94	7.75	975.97	3.39	992.975	1.71
	20	1120.3	963.79	13.97	1032.24	7.86	10.62	5.16
	21	1240.3	985.89	20.51	1076.85	13.17	1121.44	9.58
	22	1360.3	1000.95	26.41	1111.36	18.30	1170.45	13.95

Table 3.7: Forecast of Fish Weights Using 17 Data Points

Species	*Month*	*Observed*	*Model 1*		*Model 2*		*Model 3*	
			Forecast	*PE*	*Forecast*	*PE*	*Forecast*	*PE*
Rohu	18	912.3	847.52	7.10	885.19	2.97	900.90	1.24
	19	1010.3	880.09	12.88	940.46	6.91	969.10	4.07
	20	1120.3	902.10	19.47	983.17	12.24	1025.95	8.42
	21	1240.3	916.65	26.09	1015.24	18.14	1072.00	13.56
	22	1360.3	1360.3	31.91	10.38.81	23.63	1108.41	18.51

Data Collected from Fertilized Tanks

In this section we developed forecasting models using data collected from fertilized tanks and under the assumption of homogeneity of error variance. The model developed using full data is presented in Table 3.8 and models for forecasting purpose as well as validation results are presented in Table 3.9 and Table 3.10 respectively. From these results it is evident once again that the assumption of homoscedasticity is not appropriate. The validation results are very poor.

Table 3.8: Fits for Data on Rohu Under Homoscedastic Model (Full Data)

Species	*Parameters*	*Estimates*	*S.E.*
Rohu	β_1	3701.03	138.77
	β_2	-0.32	0.016
	β_3	182.43	35.64

Table 3.9: Fits for Data on Rohu Under Homoscedastic Model (Partial Data)

Species	*Parameters*	*Estimates*	*S.E.*
Rohu	β_1	2917.52	94.26
	β_2	-0.39	0.016
	β_3	308.40	55.11

Table 3.10: Forecast of Fish Weights Under Homoscedastic Model Using 19 Data Points

Species	*Month*	*Observed*	*Forecast*	*PE*
Rohu	20	2812.35	2615.11	7.01
	21	3073.35	2706.55	11.93
	22	3343.35	2771,88	17.23

Conclusion

Though forecasting fish production is very important, yet much attention is not paid to this field. Some work on forecasting is found in Walia, et al. (1977). They considered logistic models for forecasting, but the present demonstration shows that the logistic models under homogeneous error structure is not appropriate. If we consider the error structure appropriately, then any non-linear model can serve good purpose. Another point to note that the present demonstration used the data collected from cemented ponds. Naturally question may arise, how these models could be applied to natural ponds? Similar models can be developed once we have data collected from natural ponds. There are many variables such as dissolved oxygen, pH value etc., which generally affects the growth of fish. Naturally, these factors should be taken into account for developing models. Certainly these models need more complicated statistical tools and software. However, for practical purpose, the working models developed in the present demonstration may well serve the purpose.

References

Aziz, N. S., Gamberrtoglio, J. G., Lin, E. T., Grausz, H. and Benet, L. Z., 1978. Pharmacokinetics of cefamandole using a HPLC assay. *Journal of Pharmacokinetics and Biopharmaceutics*, 6:153–164.

Beal, S. L. and Sheiner, L. B., 1988. Heteroscedastic nonlinear regression. *Technometrics*, 30:327–338.

Brillinger, D. R., 1987. Comment on paper by C.R. Rao. *Statistical Science*, 2:448–250.

Davidian, M. and Giltinan, D. M., 1993. Some simple methods for estimating intra–individual variability in nonlinear mixed effects models. *Biometrics*, 49:59–73.

Draper, N. R. and Smith, H., 1981. *Applied Regression Analysis*, Wiley, New York.

Gregorie, T. G. and Schabenberger, O., 1994. Fitting bole–volume equations to spatially correlated within–tree data. *Proceeding of the 6th Annual Conference on Applied Statistics in Agriculture*, Manhattan, Kansas.

Gumpertz, M. L. and Pantula, S. G., 1992. Nonlinear regression with variance components. *Journal of the American Statistical Association*, 87:201–209.

Jenrich, R. I. and Sampson, P. F., 1976. Newton–Raphson and related algorithms for maximum likelihood variance component estimation. *Technometrics*, 18:11–17.

van Houwelinggen, J. C., 1988. Use and abuse of variance models in regression. *Biometrics*, 43:1073–1081.

Walia, S. S., Jain, R. C. and Sehdev, R. S., 1997. Non–linear statistical models for pre–harvest forecasting inland fish from ponds. Project Report, I.A.S.R.I., New Delhi–12.

Walters, D. K., Hann, D. W. and Clyde, M. A., 1985. Equations and labels predicting gross total stem volumes in cubic feet for six major conifers of southwest Oregon. Research Bulletin 50, Forest Research Laboratory, Oregon State University.

Zeger, S. L., Liang, K. and Albert, P., 1988. Models for longitudinal data : generalized estimating equation approach. *Biometrics*, 44:1049–1060.

Modelling, Forecasting, Artificial Neural Network and Expert System in Fisheries
Editors: Ajit Kumar Roy & Niranjan Sarangi
Published by: DAYA PUBLISHING HOUSE (2009)
Pages 37–46

Chapter 4

Statistical Models and their Application in Fishery Science Research

R. Venugopalan and Prajneshu

Introduction

Statistical modelling consists in constructing, a plausible symbolic representation of the functioning of a system in the form of mathematical equations. Some of the well-known models, among many others, used in fisheries are: Allometric model (describing the length-weight relationship of fish species); Growth models (modelling individual growth of a fish species); fish stock assessment models (modelling total catch as a function of fishing effort), and Multi-species models (modelling interacting/competing fish species). An excellent description of various models used in fishery science research is given by Quinn and Deriso (1999). In the following sections, we briefly describe some of the above types of fishery models.

Allometric Model

Studies of variation in growth characteristics and condition have been an important keystone in understanding early life history survival of any fish species. Change in length or weight through time have been generally used to assess the growth rates of fish populations. Some inferences drawn from such a length-weight relationship say, for example, deviations from a general length-weight relationship, *viz.* Fulton's index of condition have widely been used to describe the relative state of health of fish species. Such a relationship between fish species length and their corresponding weight is represented through a general allometric function:

$$W = aL^b \tag{1}$$

where, W is weight; L is length; *a* and *b* are parameters. The corresponding statistical model is obtained by adding an additive error term in eq. (1) (Venugopalan and Prajneshu, 1998). It may be pointed out that the resultant model is a nonlinear model. We now discuss the estimation procedures for fitting such nonlinear models.

Consider the system of equations represented by the nonlinear statistical model:

$$Y = F(\boldsymbol{\beta}) + \boldsymbol{\xi}$$

where, Z is a matrix of the independent variables, $\boldsymbol{\beta}$ is a vector of unknown parameters, $\boldsymbol{\xi}$ is the error vectors which are assumed to be i.i.d.'s following some continuous statistical distribution and F is function of the unknown parameters and the independent variables. Method of least squares is concerned with seeking a parameter vector $\boldsymbol{\beta} = (\beta_1, \beta_2, \ldots, \beta_r)$ to minimize the sum of squares

$$Q(\boldsymbol{\beta}) = \sum_{t=1}^{n} [Y_t - F_t(\boldsymbol{\beta})]^2 \tag{2}$$

To get the least squares estimates of $\boldsymbol{\beta}$, the normal equations to be solved are

$$X' F(\boldsymbol{\beta}) = X' \boldsymbol{\xi} \tag{3}$$

where, $X = \partial F / \partial \beta$. As the set of normal equations are nonlinear in nature, a closed-form solution does not exist. The proper approach in such a case is to use an iterative method in which, with a starting guess of $\boldsymbol{\beta}_0$, we try to find a new parameter estimate $\boldsymbol{\beta}$, say $\hat{\boldsymbol{\beta}}_{new}$ by minimizing the residual sum of squares $L(\boldsymbol{\beta}_0) = \boldsymbol{\xi}'\boldsymbol{\xi}$. This $\hat{\boldsymbol{\beta}}_{new}$ value is used as the starting guess for next iteration. This procedure is continued until the parameter estimates of two successive iterations yield significantly indifferent parameter estimates.

Four main methods, namely (*i*) Gradient Method, (*ii*) Gauss-Newton Method, (*iii*) Levenberg-Marquardt method, and (*iv*) Does not Use Derivatives (DUD) method are used to solve a system nonlinear equations. It is worth mentioning here that all standard statistical software packages contain programs for fitting nonlinear statistical methods using one or more of the above mentioned procedures. For example, Statistical Analysis Systems (SAS) software package has the NLIN procedure to obtain parameter estimates of a nonlinear statistical model using the Levenberg-Marquardt and DUD procedures.

Autocorrelated Errors

So far, we have studied the parameter estimation procedures to be followed for applying the allometric model with the assumption of additive error structures, assumed to be identically and independently distributed (i.i.d) with zero mean and constant variance. However, while analyzing fisheries data, like the length-weight data, which are repeated measurements made on the same individual, the above assumption is rarely met. The consequence of denial of i.i.d assumption in any regression analysis procedure is that (i) the least square estimates are no longer best (meaning those estimates which may not have a minimum variance) and consequently the estimate of error variance and the standard errors of the regression coefficients may be seriously understated, and (ii) the confidence intervals and the various tests of significance commonly employed would no longer be strictly valid. So, it is of paramount importance to check for the validity of the allometric model by examining the independency assumption of error term. We describe in the following section, the statistical test to be carried out for testing the above error assumption.

Statistical Test for Autocorrelation

The Durbin-Watson test is a very popular test of autocorrelation in regression analysis. The test is based on the assumption that the errors constitute a first-order autoregressive series, namely

$$\xi_t = \rho \xi_{t-1} + u_t, \; |\rho| < 1$$

where, u_t are independently and normally distributed with zero mean and constant variance and p is the correlation coefficient between successive residuals. The test statistic d is defined as

$$d = \sum_{t=2}^{n} \left(\xi_t - \xi_{t-1}\right)^2 \Big/ \sum_{t=1}^{n} \xi_t^2$$

and is utilized for testing $H_0 : p = 0$ against $H_1 : p > O$. The parameter p is estimated by r,

where, $r = \sum_{t=2}^{n} \xi_t \xi_{t-1} \Big/ \sum_{t=1}^{n} \xi_{t-1}^2, \; -1 \le r \le 1, \; d \approx 2(1-r)$

Evidently, d has a range from 0 to 4. Since r is an estimate of p, it is clear that d is close to 2 when $p = 0$; near to zero when $p = 1$. Hence, it is inferred that evidence of autocorrelation among observations is indicated by the deviation of d value from the numerical value of 2. We consider the following decision rules for testing the hypothesis of absence of autocorrelation among the residuals. If

(*i*) $d < d_L$, reject H_0

(*ii*) $d > d_u$, do not reject H_0

(*iii*) $d_L < d < d_u$, the test is inconclusive.

The values of (d_L, d_u) are respectively the lower and upper limit values at α per cent level of significance corresponding to n degrees of freedom. A possible rejection of hypothesis prompts us to fit the allometric model with auto-correlated error structure as given by eq. (1). A commonly employed technique for parameter estimation in the case of nonlinear regression with auto-correlated error structure is a "Two-stage" estimation procedure, which could be carried out using SAS software package.

Growth Models

Processes producing sigmoidal or "S-shaped" growth curves are widespread in the field of agriculture, biology and fisheries. The underlying mechanisms of these growth curves are that they start at some fixed point and increase their growth rate monotonically to reach an inflexion point; after this the growth rate decreases to approach asymptotically some final value. In fisheries, as the fish tend to grow continuously throughout their life in areas where there are marked seasonal differences in the environment (food, temperature, etc.), most of the fishery biologists will be interested in finding a curve describing such a fluctuating pattern of growth satisfactorily. Accordingly, a number of growth models such as Schnute' s versatile model, Von-Bertalanffy model, Richards growth, Gompertz growth, logistic growth and linear growth models are developed in the fishery literature. We now discuss these models briefly:

Denoting the Size of a fish by W, the rate of growth in a fish is expressed as a derivative

$$Z = \frac{1}{W}\frac{dW}{dt} \tag{4}$$

Just as one might study how the size develops in relations to time, a fishery biologist might also examine the development of the relative growth rate $Z(t)$.

Considering the simplest possible assumption that the relative growth rate of the relative growth, i. e. $(1/Z)(dZ/dt)$ is a linear function of $Z(t)$

$$(1/Z)(dZ/dt) = -(k + bZ) \tag{5}$$

where, *k* and *b* are constants. By solving the pair of differential equations (1.2) and (1.3) with suitable end conditions, provided a versatile growth model (hereafter called as the Schnute growth model) for different values of *k* and *b*. These are represented by equations (6)–(9):

Case (i): When $k \neq 0, b \neq 0$,

$$W(t) = \left[w_1^b + \left(w_2^b - w_1^b\right) \frac{1-e^{-k(t-t_2)}}{1-e^{-k(t_2-t_1)}} \right]^{1/b} \tag{6}$$

Case (ii): When $k \neq 0$, b=0,

$$W(t) = w_1 \exp\left[\log(w_1/w_2) \frac{1-e^{-k(t-t_2)}}{1-e^{-k(t_2-t_1)}} \right] \tag{7}$$

Case (iii): When k = *0*, b ≠ 0,

$$W(t) = \left[w_1^b + \left(w_2^b - w_1^b \right) \frac{(t - t_1)}{(t_1 - t_2)} \right]^{1/b} \tag{8}$$

Case (iv): When k = *0*, b=0,

$$W(t) = w_1 \exp\left[\log(w_1 / w_2) \frac{(t - t_1)}{(t_2 - t_1)} \right] \tag{9}$$

In the above equations, w_1 and w_2 denote the sizes of a fish at two particular ages in its life span, say t_1 and t_2. Further, it is assumed that $w_1 > w_2 > 0$ and $t_2 > t_f$. The major advantage of the Schnute growth model is that for different suitably chosen values of *k* and *b*, the commonly available historic growth models are obtained (see Table as below)

Special Cases of Schnute Growth Model

Sl.No.	Choice of k, b	Model title	Corresponding Equation Number
1.	K > 0, b > 0	Von- Bertalanffy growth	(10)
2.	K > 0, b = 3	Putter curve	(11)
3.	K > 0, b = 0	Gompertz growth	(12)
4.	K > 0, b < 0	Richards growth	(13)
5.	K > 0, b = –1	Logistic growth	(14)
6.	K = 0, b = 1	Linear growth	(15)

Von-Bertalanffy Growth

For $k > 0$ and $b > 0$, eq. (6) results in the most widely used Von-Bertalanffy growth (VBG) model in fisheries (Venugopalan and Prajneshu, 1997):

$$W(t) = w_\infty \left[1 - e^{-k(t - t_0)} \right]^{1/b} \tag{10}$$

Putter Curve

This is a specialized VBG with $b = 3$, represented as

$$W(t) = w_\infty \left[1 - e^{-k(t - t_0)} \right]^{1/3} \tag{11}$$

Gompertz Growth

For $k > 0$, eq. (7) boils down to the Gompertz model,

$$W(t) = w_\infty \exp\left[-\exp\left\{ -k(t - t_0) \right\} \right] \tag{12}$$

Richards Growth

Taking $k > 0$ and $b < 0$, eq. (6) reduces to the Richards curve represented by

$$W(t)=w_{\infty}\left[1+b e^{-k(t-t_0)}\right]^{1/b} \tag{13}$$

Logistic Growth

This is a specialized Richards model with $b=1$, written as

$$W(t)=w_{\infty}\left[1+e^{-k(t-t_0)}\right]^{-1} \tag{14}$$

Linear Growth

For $k = 0$ and $b = 1$, eq. (8) reduces to

$$w(t)=k(t-t_0) \tag{15}$$

In all the above growth models, the unknown parameters, *viz.* w_{∞}, k, b have important biological interpretations. Specifically, the parameter $w_{\infty} = \lim W(t)$ represents the asymptotic size; k represents the constant relative rate of relative growth rate of a fish; b refers to the incremental relative rate of relative growth rate of the fish and $t_0 = W(t) = 0$ corresponds to age of theoretical zero size, which also represents time when the growth curve crosses the t-axis. It is also worth highlighting that small fast growing fish species are characterized by large k and small w_{∞} values. At the other end of the spectrum, long-lived slow growing species have small k values and larger w_{∞} values.

Fish Stock Assessment Models

Stock assessment in fisheries is basically aimed to advice on the optimum exploitation of aquatic living resources. Living resources are limited but renewable, and assessment of such a stock is essentially made to search for the exploitation level, which in the long run, gives the maximum yield in weight from fishery. To carry out such a stock assessment in any fisheries, two main groups of fish stock assessment models extensively used are: (i) Holistic models and (ii) Analytical models. The basic difference between these two types of models is that the former employs fewer parameters than the latter. The type of model to be used depends on the quality and quantity of input data available with the data analyst. The analytical models, which are believed to give more reliable prediction demand detailed description of the stock. On the other hand, the data analyst may prefer to use holistic models whenever there is not adequate information regarding the stock. We now briefly discuss some of the well-known stock assessment models of each type, *viz.* holistic models and analytical models, available in the fishery literature.

Holistic Models

Holistic models in fish stock assessment consider a fish stock as a homogeneous biomass and do not take into account, for example, the length or age structure of the stock. One of the well-known models of this kind is the 'Surplus production models'. The objective of the application of such models is to determine the optimum level of effort, that is the effort that produces the maximum yield that can be sustained without affecting the long-term productivity of the stock, the so-called Maximum sustainable yield (MSY).

Surplus production models, as developed, facilitate estimation of MSY and the optimum fishing effort for harvesting the MSY (f_{MSY}). If B_t denotes the biomass of a stock at time t, the Schaefer model is described by the differential equation: $dB_t/dt = rB_t(1 - B_t/k) - F_t B_t$

where, r is the intrinsic growth rate; K is the carrying capacity; and $F_t = qf_t$ is the fishing mortality at time t defined as the product of constant catch ability coefficient and the fishing effort at time t. Similarly, if B_t denotes the biomass of a stock at time t, the Fox model is described by the differential equation: $dB_t/dt = aB_t - bB_t \log B_t - F_t B_t$

Another important series of surplus-production models is Pella-Tomlinson model (Venugopalan and Prajneshu, 2000). This model includes both the above models as special cases but at the expanse of an additional parameter (m) and is given by the differential equation: $dB_t/dt = aB_t - bB_t^m - Y$

The utility of the above models lies is obtaining MSY and the corresponding effort to achieve the MSY (f_{MSY}). Accordingly, under the equilibrium condition of fish population, MSY and f_{MSY} values of the various surplus production models are obtained as

$$MSY_{Schaefer} = a^2/(4b),\ f_{MSY} = a/(2b) \tag{16}$$

$$MSY_{Schaefer} = -\exp(a-1)/b,\ f_{MSY} = 1/b \tag{17}$$

$$MSY_{Pella-Tomlinson} = (1/b)(1-1/m)(1/m)^{1/(m-1)}\, a^{m/(m-1)},$$

$$f_{MSY} = (a/b)(m-1)/m \tag{18}$$

where, $a = qk$ and $b = q^2 k/r$.

Non-equilibrium Version of Schaefer Model

Aquatic living resources are limited but renewable and fish stock assessment may be described as a search for the exploitation level which, in long run, gives maximum yield from a fishery. To this end, surplus fish production models, because of their simplicity and relatively undemanding data needs, are attractive tools. Among the class of such models, Schaefer model, as described earlier, has been extensively used. Unfortunately, researchers in the past have mostly confined their attention to applying only the equilibrium version of this model. However, it is well recognized that fish stocks are rarely, if ever, in equilibrium. Prajneshu and Venugopalan (1998) evaluated comparative performance of equilibrium vis-à-vis non-equilibrium Schaefer model for estimation of MSY and related parameters. As results obtained by two approaches were found to differ markedly, it was concluded that non-equilibrium approach should invariably be adopted in order to capture reality.

It may be worthwhile to mention here that, in general, information obtained about biological quantities using a single set of time-series data about fish stock during a particular time–period is utilized to assess status of the stock. However, in doing so, care must be taken to take into account bias arising due to sampling a particular data set. Hence, there is a need to provide estimates of biological quantities together with a range in which these are expected to fall, if different sets of data of the same stock were to be considered. This shall undoubtedly strengthen inference about the stock. This may also help the policy makers to suggest a flexible management strategy to take into account the above said sampling bias. To this end, bootstrap technique may be employed. Bootstrap bias-corrected confidence-intervals (C.I.) are constructed for non-equilibrium Schaefer model to assess reliability of parameter estimates using simulated data sets.

Bootstrap technique is a sample reuse technique often used to estimate sampling variances, confidence-intervals, bias, and similar properties of statistics, including parameter estimates. It is often performed by resampling original observations, which are generally unordered. However, in fitting non-equilibrium surplus production models to time-series observations, order of catch-effort pairs is as important as original data themselves. Accordingly, to preserve original time structure of data, following bootstrapping method is applied:

Parameter estimates. Denoting by $\hat{f}$, the vector of ordered predicted efforts obtained by fitting non-equilibrium version of Schaefer model for catch-effort data series, a set of synthetic values, say f^* are constructed for each bootstrap trial by combining ordered predictions from original fit with a randomly chosen residual (with replacement) from a set of residuals from original fit. The model is then refit, using pseudo-values in place of original observed values of effort. This process is repeated for 1000 such bootstrap trials. Using the results obtained, median estimation bias $B_{\theta,M}$ in a parameter θ is

$$\hat{B}_{\theta,M} = \hat{\theta}_M - \theta, \tag{19}$$

where, $\hat{\theta}$ is conventional estimator of parameter θ, and $\hat{\theta}_M$ is median value of θ obtained from bootstrap trials. A bias-corrected estimator $\hat{\theta}_{BC}$ BC of parameter θ is given as

$$\hat{\theta}_{BC} = \hat{\theta} - \hat{B}_{\theta,M} \tag{20}$$

Construction of confidence-intervals. Bias-corrected (BC) method, the most widely followed confidence-interval construction procedure, is used to construct confidence intervals and is described below:

Let $F(Z)$ be distribution function (DF) of standard normal distribution and let Φ^{-1} be its inverse. Let $\hat{C}$ be empirical bootstrap DF of parameter θ, *i.e.* $\hat{C}(g)$ is proportion of realizations of $\hat{\theta}$ in bootstrap distribution that falls below any arbitrary value g. Define the constant

$$Z_0 = \Phi^{-1}\left[\hat{C}\left(\hat{\theta}\right)\right],$$

where, $\hat{\theta}$ is conventional estimator. Then, 100(1 - 2α) per cent BC central confidence-interval for q is defined as

$$\left\{\hat{C}^{-1}\left[\Phi\left(2Z_0 + \Phi^{-1}(\alpha)\right)\right]\right\}, \left\{\hat{C}^{-1}\left[\Phi\left(2Z_0 - \Phi^{-1}(\alpha)\right)\right]\right\} \quad (21)$$

In present communication, using the procedure described above, median estimator and BC confidence-intervals are constructed for the main biological quantities such as MSY, corresponding effort to achieve this MSY (f_{MSY}), growth parameter (r), carrying capacity (K), fishing mortality rate at MSY (f_{MSY}), stock biomass level at which MSY is materialized (B_{MSY}), and also two ratios, *viz.* B-ratio and F-ratio. A statistic, called as Relative inter-quartile range (RIR) is also calculated by dividing bias-corrected 50 per cent confidence-interval (inter-quartile range) by corresponding median bias-corrected estimate. Resultant statistic may be viewed as a non-parametric analogue of coefficient of variation. Lower the values of RIR, more accurate are estimates of above quantities.

Analytical models

Some of the well-known analytical models available in fishery literature are:

Beverton–Holt model

Let $B_S(t)$ denotes the biomass of the fish population belonging to the sth cohort at time *1*. Under the assumptions: (a) the sth cohort enter the fishery (recruits) at time $t = s$ and that the; recruitment R is a constant which is same for all cohorts, (b) the fishing gear captures all fish of age $t \geq s + t\mu$ (where is μ is mesh-size) and all small fish escape, the general form of the Beverton-Holt model is given by the differential equation:

$$\frac{dB}{dt} = \begin{cases} -MB_S, & s \leq t \leq s + t_\mu \\ -(M+F)B_S, & t \geq s + t_\mu \end{cases} \quad (22)$$

where, M is the natural mortality and F is the fishing mortality. The disturbing feature of this model as compared to holistic ones is that it requires detailed data on mortality, age structure and growth rate. Hence it is seldom used for stock assessment.

Thompson and Bell Model

This is an age-based predictive model used to predict the effect of changes in the fishing effort on future yield and also the biomass levels. This provides a very important tool for the fishery biologist to demonstrate the effect that certain management measures, such as change in the minimum mesh size, decrease or increase of fishing effort, or closed seasons will have on the biomass and the value of the catch.

Conclusion

The important role played by a variety of statistical models in fishery science is highlighted. It is hoped that, in future, these will find more applications to India's fisheries data.

References

Prajneshu and Venugopalan, R., 1998. Comparative study of nonequilibrium vis-à-vis equilibrium Schaefer model. *Ind. J. Fish.*, 45:1-6.

Quinn, T. J., II and Deriso, R. B., 1999. *Quantitative Fish Dynamics*. Oxford University Press, New York

Venugopalan, R. and Prajneshu, 1997. Von Bertalanffy growth model with auto-correlated errors. *Ind. J. Fish.*, 44:63-67.

Venugopalan, R. and Prajneshu, 1998. On nonlinear procedure for obtaining length-weight relationship. *Ind. J. Ani. Sci.*, 68:70-72.

Venugopalan, R. and Prajneshu, 2000. Pella-Tomlinson nonlinear statistical model with auto-correlated errors. *J. Ind. Soc. Ag. Stat.*, 53:12-19.

Modelling, Forecasting, Artificial Neural Network and Expert System in Fisheries *Pages* **47–66**
Editors: **Ajit Kumar Roy & Niranjan Sarangi**
Published by: **DAYA PUBLISHING HOUSE (2009)**

Chapter 5

Methods of Forecasting and their Application in Fishery Science

P.K. Sahu

Ever since the initiation of global battle against the vicious circle of poverty, food insecurity and hunger, agricultural statistics is playing a great role. Reliable and accurate food and agriculture statistics is providing immeasurable support to this global battle. In this process, forecasting of food and agriculture parameters can contribute in minimizing the disaster from food insecurity and other natural calamities, particularly in those countries, where agriculture is the backbone of national economy. In most of the developing countries like India, Pakistan, Bangladesh, Ethiopia, Sri Lanka etc. a considerable percentage of population is directly or indirectly dependent on agriculture and agriculture related activities for their daily livelihood. A substantial contribution to nation's gross domestic product is coming out of agricultural activities. For example, for the triennium 2001-02 to 2003-04, in India agriculture contributed 22.6 per cent to the national output (Ramesh Chand, 2005). It is fact that from a food deficient country, India has turned itself into a food surplus country of 60 million tones of rice and wheat in buffer to meet any natural calamity. On the other hand it is also reported that over 200 millions of its population going to bed everyday either fully or partially hungry (Siddiq, 2002). If this section of the people gets an access to food then it is a matter of great concern whether this buffer stock will remain at all or not! Among the various agricultural sectors, aquaculture is one of the fastest growing sectors with continuous average growth rate of over 6 per cent for the last twenty years or so. Now the question is even with this growth rate of 6 per cent and average productivity of 2.4t/ha can aquaculture be efficient to meet the future demand for fish and fish product in India and abroad? Along with this, the ever-increasing population pressure has also made this battle more difficult against food insecurity and poverty. Brugere and Ridler (2004) in an attempt to forecast the national aquaculture productions of different countries using three different forecast models by Delgado et al., 2003; Wijkstrom,2003 and Ye, 1999 suggested for a planning frame work for issues to address with technique like Delphi method to improve future plan and likelihood success of the plans. With this pre-text, it becomes essential to bring together the different techniques of forecasting which are being used or could be used in aquaculture for forecasting purpose.

Agriculture and its allied sectors are unique in the sense that its policy being framed at one level (official/government/institutional) whereas its execution takes place at the micro level (*i.e.* farm). Like other sectors of agriculture, for successful implementation of aquaculture improvement policies, different components are to be integrated together. To win over the battle against food and nutritional insecurity through sustainable enhanced production from aquaculture sector one has to take in to consideration of the following aspects:

(*a*) Prediction of likely population

(*b*) Prediction of per capita requirement of fish and fish products for the population in (*a*)

(*c*) Prediction of productivities and production from different components of fish production system.

(*d*) Forecasting of weather and climate

(*e*) Forecasting of energy demand in aquaculture.

(*f*) Forecasting of input and feed requirement in different seasons for different fish production components.

(*g*) Forecasting of demand and supply of inputs and feed.

(*h*) Forecasting of disease and pest incidence

(*i*) Forecasting price and market demand for different fishery products.

(*j*) Forecasting of trained manpower required for the successful implementation of aquaculture policies.

The most challenging task to the agriculture planners is to provide food and nutritional security to the present population as well as to plan for supply of food to the future population of the country. Therefore, the planners should have clear idea about the possible future population and accordingly their food and nutritional requirement. Projection of likely population is being done by various agencies at national and international level. At national level in India, National Sample Survey Organisation (NSSO), Central Statistical Organisation (CSO) etc are the main agencies. Where as at international level different bodies of United Nations (UN) like International Labour Organisation (ILO), World Health Organisation (WHO), Food and Agriculture Organisation (FAO) are doing the job. The process is to use the time series data and combine with certain population related indicators to get the projection. Any faulty prediction of population is bound to affect the subsequent planning at different steps. Once decided the food requirement, the next task is to have idea on food habit of people at different parts of the country. Demand for food fish is influenced by various factors like demography, living standard, urbanization, price of fish and fish products, prices of alternative animal proteins etc. In countries like China and India though the population growth rates have declined but there is a trend of rising per capita real income. So, one can expect an income–induced demand combined with urbanization towards more demand for food fish in these countries. Per capita consumption of fish is an important factor in estimating demand for food fish. Wijkstrom, 2003 has suggested that though fish has become more important in people's diet, it contributes only 15.3 per cent in total animal protein taken by the human beings and there is potential for declining marginal utility and consumer saturation as income rises. While making projection for productivities of different food items using forecasting technique, one has to take care of the state of technology available and would be available during the period of projection. Estimating the productivities of different components of fishery, steps will be to project the quantum of land and water area available along with their diversity and the state of development under going in these areas. Is the area too large or too small? Is the area homogeneous or diversified? Has the area (land and water bodies) designated for specific purposes? Is the current state of development sufficient to meet the demand or appropriate correction measure to be taken up in accordance with the difference in forecasted requirement and forecasted realization?

Weather and climate are the key factors in aquaculture. The ever-increasing pressure on land due to population influx, urbanization, industrialization is putting tremendous pressure on environment. Different legislations are coming into force to protect the environment which are to be taken into consideration while dealing with aquaculture activities with particular reference to coast line aquaculture. Forecasting of weather and climate is getting tremendous importance not only because of its importance to aquaculture but also for its wide utility in various fields. Forecasting of weather is mainly of three types a) short range weather forecasting, b) medium range weather forecasting and c) long term weather forecasting. Accuracy of long range forecasting in India though satisfactory, medium and short range forecasting are improving with the active support from the forecasters. Taking the experience of failure in predicting the short fall in 'Belg' (short season) rainfall and deficiency in 'Kiremt' (main season) rainfall resulting drought in 1984, Ethiopia could develop forecast method to predict the possible *El nino* situation in 1989, is an example how forecasting can help in managing disaster.

The success of aquaculture substantially depends on the timely availability of feed and other inputs and their prices. Under the present free world trade regime forecasting of demand–supply and prices of feed and other inputs greatly influences the success of the aquaculture production system, as such forecasting has to play significant role in this aspect.

Yield loss in crop agriculture due to pest is estimated to about 30 per cent and as such disease and pest management is one of the vital components of agricultural management. In aquaculture also, forecasting of disease and pest and thereby possible loss in production plays vital role in meeting the requirement. De Silva, 2001 reported about the vulnerability of the shrimp production towards diseases. During 1990's there was tremendous disease threat in the shrimp production of Equador. Although many microbial and viral infections are not considered to be direct threat to human health, definitely a threat to marketability of food fish; particularly under the present world free market scenario emphasizing on quality product. As it is well known that incidence of disease and pest has significant relationship with the weather components, so forecasting of disease and pest taking care of possible weather and climate is of great importance in terms of reducing the loss due to disease and pest.

Accurate forecasting of fish production under different components can help in planning the production process for seasons to follow in order to overcome any shortfall or surplus productions over planned production. For example, Roger, 2005 has reported the use of abundance of juvenile brown shrimp and different environmental variables like temperature, rainfall, bay water height etc. to assess the favourable condition for growth and survival of young shrimps towards forecasting the availability of brown shrimps. Use of Baxter Bait Index versus offshore landing regression model for past years resulted a shortfall of 3 million pounds of shrimps during 2005 in Texas offshore waters.

Energy demand forecast is another area in which forecasting play a major role. Planning for forecasted percentage increase in energy consumption due to increase in percentage of forecasted production should be done with utmost sincerity; otherwise, there will be difficulty in achieving forecasted crop production. Previously, forecasting energy demand was largely based on aggregate macro level data and overall national average and trends. These forecasts were unable to capture variations between components, regions, operations and level of technology used. The energy demand forecast should give a picture to percentage increased in energy consumption required to achieve 1 per cent increase in production.

To complete all the process for development, a country need to have adequate trained manpower in harnessing the profits from forecasting at every step. Therefore, forecasting for the requirement of trained or different types of manpower required for aquaculture development should be done in such a way so as to maximize the forecasting accuracy and meet the requirement of manpower at different stages of aquaculture development.

The above enumeration is not exhaustive one, there are certain other areas of aquaculture system where the art and technique of forecasting can play major role in augmenting the aquaculture production and other inter related aspects and ultimately to win over the battle against poverty and hunger and nutritional insecurity, besides economic solvency.

Different forecasting methods applied in aquaculture can broadly be classified under long range, medium range and short range forecasting. Among the different methods used in forecasting (*a*) Expert opinion, (*b*) Delphi method, (*c*) PRA technique, (*d*) Statistical methods, (*e*) Analogue method, (*f*) Use of remote sensing information, (*g*) Neural network etc., are important and widely used, of course this is not an exhaustive list of methods!

Expert Opinion Method

Expert opinion method is the age-old method and still being practiced. In this method, opinion(s) from the expert(s) in a particular field is sought and decision on consensus basis is arrived at. This method is highly subjective in nature and depends on the up-to-date information gathering and analyzing capacity of the expert(s). Forecasting in this process is dependent on the perception, beliefs, wisdom, knowledge and experience of the experts. In aquaculture, old and experienced fishermen are generally recognised as experts. Sometimes it becomes very difficult to justify the reason behind a particular forecast. As such this method may or may not be statistically sound.

Delphi Method

One-step ahead of expert opinion is the 'Delphi method'. The Delphi method is a very strong structured process for collecting and distilling knowledge from a group of experts by means of a series of questionnaires interspersed with controlled opinion feedback, has mostly been used as mono-variable exploration technique to generate forecast in technology, education etc. This method recognizes human judgment as legitimate and useful inputs in generating forecasts. The success of Delphi method depends on a) kind of group of communications process available in hand to explore the problem, b) kind of experts and locations of experts and c) alternative technique available. Failure to address these aspects may lead to inappropriate application of the method. The most striking feature of this method is that the feedback of the experts is summarized statistically rather than in terms of majority vote. This method has wide applications in various fields of agriculture where there is no or little information available. Gotsch and Reider (1990) have reported the use of Delphi method to forecast the future development of crops protection in agriculture. There are several other examples where Delphi method has been used for forecasting purpose. Like expert opinion method this method also is based on the judgment capability and forecasting capability of the experts in panel. Moreover, there may be a tendency on the part of the expert to discount the future as well as the past along with simplification urge, resulting in inappropriate or inaccurate forecasting.

PRA and RRA Method

Participatory Rural Appraisal (PRA) and Rapid Rural Appraisal (RRA) techniques are the processes in which forecasting values are generated for immediate and short-term use. People share their views on particular aspect and a consensus is arrived at. It has region and time specific interest. In most of the time, it is found that adoption and sustainability of technically feasible package for improvement of aquaculture is slow or limited because of some non-technical factors, which planners fail to take in to account. PRA or RRA methodologies are very much efficient to overcome these problems. These techniques bridge the gap between technical planning and the grass root level reality. Both these method depend on multidisciplinary, integrated approach to developmental issues. These are useful when they are fed information and learning in to planning and forecasting system which are interdisciplinary and integrated by themselves. These are cost effective, rapid and encourage community level developmental initiative at the same time may generate aspirations and expectations from the community, but are very much effective in increasing forecasting accuracy based on ground truth. These methods are also not/ little subjected to statistical methodologies.

Analogue Method

Analogue method is based on assumptions that a current synoptic situation will likely develop in the same way as similar past synoptic situations (WMO, 1992). The efficiency of this method is mostly dependent on the proper selection of the past analogue year. This process has been used in rainfall forecasting in Ethiopia and accordingly a year is categorized as above normal (>125 per cent), normal (75 per cent–125 per cent), below normal (50 per cent–75 per cent) and much below normal (<50 per cent). In early days the expert fishermen used this technique for fishing purpose either knowingly or unknowingly. Problem with this method is that no two years are identical and hence the selection of analogue year.

Judgmental Forecasting

Once judgment was thought to be hindrance to the forecasting accuracy, today judgment has been recognized as an indispensable component of forecasting and being used to improve forecasting efficiency. During the last two or three decades lot of researches have been done on integration of judgmental forecasting with other methods of forecasting. Judgmental forecasting is also subjected to bias, without management, judgment in forecasting may result serious problems. A survey on 240 US corporations (Sender and Manrodt, 2003) revealed that only 11 per cent forecasting are solely dependent on forecasting software while 60 per cent adjust forecast based on their judgment.

After getting the data on time series or cross-section, forecasting is made through different methods of forecasting using different forecasting softwares. Time series data are sometimes referred to as history data and the other one as the domain or contextual data. The domain data are in-fact used to help understand the past and project the future. In practice majority of the situations, judgmental forecasters will have some information about the value to be forecasted in addition to the time series data. Lawrence, Edmundson and O'Conor (1985) first compared the accuracy of judgmental forecasting and quantitative model forecasting in a large scale using real life data. They concluded that the judgmental forecasting to be at least as accurate as statistical techniques, besides in a number of cases the judgmental technique as the most accurate. Bolger and Wright (1994) demonstrated that the performance of judgmental forecasting is a function of the interaction between the dimension of ecological validity and learnability.

The use and understanding of domain knowledge is very much important for accuracy of judgmental forecasting. It is our common experience that a forecaster gets both the time series data as well as the contextual or domain knowledge about the time series data which not only helps in understanding the past behaviour of the series but also predicting the future. The most characteristic feature of domain knowledge is that it is the un-modeled component of the time series and if it is being modeled by a statistical method, it can be incorporated into the statistical forecast itself. The most suitable situation where judgmental forecasting can be exploited is a situation where both contextual knowledge (knowledge of the nature of the time series) and some additional irregular knowledge that can be useful in either explaining the past behaviour of the series or in predicting the future (or both) are available.

The main problem with the judgmental adjustment in forecasting is that a forecaster analyses the pattern of noise associated with the series. Obviously, the forecaster adjusts the statistical forecast that he supposed to miss in the pattern, at the cost of forecast accuracy. However, adjustment can even improve the accuracy of statistical forecast under right condition. In cases where the statistical method is deficient, to some extent, in estimating pattern of the time series, judgmental forecasting may improve the accuracy. Similarly judgement can make an improvement of the statistical forecasting where the forecasters have got important domain knowledge, that has not been included or available to the statistical method. The efficacy of adjustment depends on whether a statistical forecast needs adjustment and the size or quantum of adjustment is required. Thus, judgement can play an important role in forecasting if it is applied appropriately.

Statistical Forecasting

Most widely used technique is the use of statistical methodology. Under this broad methodology, there are uses of several statistical techniques/tools, among these, use of regression, probability, parametric methods, non-parametric methods, semi-parametric methods etc. are some of the important methods used in forecasting, definitely, the list is not exclusive one. Time series modelling has also come in great way in to the arena of agricultural forecasting. Among the time series models time-trend model, the autoregressive model (AR), the autoregressive moving average model (ARMA), the autoregressive integrated moving average (ARIMA) model, the autoregressive conditional heteroscedastic (ARCH) model, the generalized autoregressive conditional heteroscedastic (GARCH) model, are important because of their wide use in forecasting. Certain non-parametric models like kernel, loess; semi-parametric models like spline has also been used to bypass the assumptions of parametric modelling. The non-parametric/semi-parametric methods put great emphasis on data points and as such are of great importance in real life data situation (Pal and Sahu, 2004). There are several examples where these methods have been used successfully with particular reference to agriculture. The simplest among the forecasting technique is the regression technique. This method has been used either in isolation or in combination with other methods. For example, Brazil *et al.* (2006) in an attempt to forecast the togiak herrings population in 2007 have used age structured analysis model with the help of catch and age composition data and aerial survey biomass estimates. Simple linear regression models were used to forecast the weight of age-5 through age-15.

Time Series Forecasting

Time series forecasting occupies a major part of the total forecasting world. Many works have been done on forecasting with time series data. This is because of the fact that time series data are comparatively easy to be available if one goes through the past records. Almost all standard methods (namely exponential smoothing, Box-Jenkins methods of modelling, Non-linear models like Regime-switching models,

Functional co-efficient models, Neural-Network model, Long-memory model, ARCH-GARCH model etc. either in isolation or in combination) of forecasting have been used in time series forecasting.

Exponential Smoothing

The method of exponential smoothing though started during 1950s received a boost during late 80s, after the publication of the papers by Gardner (1985) and Snyder (1985). The simple exponential smoothing (SES) method is based on averaging the series data in exponential (decreasing) manner. The weights used for the purpose are w, w(1-w), $w(1-w)^2$ ..... for the most recent period t, (t-1), (t-2) ...... respectively. The weight w is known as the smoothing parameter or smoothing constant and its value ranges between 0-1. Through SES method, we generally obtain the following type of equation:

$$S_t = wY_t + w(1-w)\,Y_{t-1} + w(1-w)^2\,Y_{t-2} + \dots\dots\dots$$

$= wY_t + S_{t-1}$. where, S_t and Y_t are the smoothed value and actual value respectively for the year t and S_t is used to forecast the value for the period (t+1). Generally in estimating the smoothing parameter w, we minimize the mean square error (MSE) defined as

$\frac{1}{n}\sum_{t=1}^{n}(\hat{Y}_t - Y_t)^2$, where $\hat{Y}_t$ and Y_t are the estimated and observed values respectively corresponding to the period t.

In addition to its extensive use in business and industry, it has profuse application in areas dealing with state-space model in particular. The method is being treated as robust because of its remarkably good forecasting performance; sometimes the simple exponential smoothing (SES) perform even better than first order ARIMA model. In agriculture, SES has been used (Pal and Sahu, 2003) on pest forecasting of a particular crop taking time series data of pest infestation. Similar use can also be made in studying the catch behaviour in aquaculture over a long period of time.

Box-Jenkins Method of Forecasting

Yule (1927) initiated the first step towards the formation of auto-regressive modelling. The idea was to regard every time series as the realization of stochastic process. Later on researchers like Walker, Slutsky, Yule and others conceptualized the auto-regressive (AR), moving average (MA) models. With the publications of the "Time-series analysis: forecasting and control" by Box and Jenkins in 1970 there was tremendous boosting in the study of time series forecasting. Because of which the technique of integrating, the identification of cycles, estimating the parameters of the models and verification of the forecasting model based on time series data is popularly termed, as Box-Jenkins's method (B-J method).The use of this technique was still restricted because of large amount of computational requirement of the method. However, the introduction of computers and software further boosted the use of Auto Regressive Moving Average (ARMA), Auto Regressive Integrated Moving Average (ARIMA) models, multi-variate vector ARIMA models in forecasting. As because a number of competitive models can be framed with the help of B-J methods from a to the researchers. Many techniques or methods have been suggested in this endeavor, among these Akike's information criterion (AIC), Akike's final prediction error (FPE) and the Bayes information criterion (BIC) are widely used for the purpose.

A number of methods are there in estimating the parameters of B-J models. Most of the methods, though asymptotically equivalent, sometimes there are large differences in finite sample properties. Newbold *et al.* (1994) recommended the use of full maximum likelihood method for estimation of parameters. Recently, in the year 2003, Kim demonstrated that bootstrap bias corrected parameter estimators giving rise to more accurate forecast than the least square estimators did.

The univariate ARIMA modelling technique has been generalized to multivariate cases-known as vector ARIMA (VARIMA). Like AR, ARMA, ARIMA models in multi-variate cases there are vector auto-regressive series (VAR), vector auto-regressive moving average (VARMA) models, and vector auto-regressive integrated moving average (VARIMA) models. B-J method of time series forecasting both in uni-variate and multi-variate cases have been extensively used in labour market forecasting, business forecasting etc. The concept of co-integration has raised the question on forecasting ability of the error correction models (ECM) over unrestricted VARs. Empirical studies (Wang and Bessler, 2004) suggested that ECMs outperform VARs in longer forecast horizons.

Forecasting of different economic, agricultural variables has been made using the B-J method of forecasting. This method of forecasting can very well be used in aquaculture production and its allied parameters for proper planning in the development of these sectors. Parsons and Colbourne (2000) have used ARIMA modelling technique to get predictive model for shrimp catch per unit effort (CPUE) in fishing area off the mid-Labrador coast.

Non-linear Models of Forecasting

The assumption of linearity of model, though very useful and powerful tools, in many cases, it was felt increasingly that linear models are not suitable for application in many real life situations. The major problem with the non-linear model is the tremendous computational work compared to linear models. With the advent of computer technology and software, more and more non-linear models are being practiced for the purpose of forecasting. Now-a-days non-linear forecasts are obtained by either Monte Carlo Simulation or by bootstrapping method. Bootstrapping method has the edge over the Monte Carlo Simulation because of no assumption about the distribution of the errors.

The extensive use of piece wise linear models is found in the class of threshold auto-regressive models, which are self-exciting. Thus, the self-exciting threshold auto-regressive (SETAR) models are essentially piece wise linear model in basic form. According to Clements and Smith (1997), Monte Carlo Simulations are satisfactory when the disturbances in the SETAR model have a symmetric distribution; otherwise, the bootstrap method is good. One draw back of the SETAR model is that the dynamic changes discontinuously from one regime to another regime. To overcome this, a new type of smooth transition auto-regressive (STAR) model was proposed and evidences was put forwarded by Sarantis (2001) to have improved forecasting models. However, Bradly and Jensen (2004) commented against the above evidence of Sarantis.

In contrast to simple auto-regressive models, the functional coefficient auto-regressive (FCAR) model is a model in which the auto-regressive model coefficient are treated as the smooth functions of lagged time value of the series or an exogenous variables. The FCAR model incidentally includes the SETAR, SITAR, and STAR models in special cases. Harvill and Ray (2005) have presented a more efficient forecasting model using univariate and multi-variate functional coefficient (V) FCAR models.

Most of the non-linear models have the universal approximation ability. Barron (1993) showed that this ability of universal approximation is of much interest for the functions defined by the neural networks. This ability allows non-linear models to outperform linear models if there exists non-linear relationship between the variables. Artificial neural network (ANN) is another useful tools in forecasting, particularly when the functional relationship is very difficult to read or unknown. In ANN process, the variables (inputs and dependent variables) are filtered through some hidden layers having hidden units or nodes before the output is obtained. Virtually these intermediate outputs are the linkage between the inputs and the ultimate results. Doraiswami *et al.* (2003) has predicted crop yield in north Dakota using simulation technique with remote sensing data for different crop growth phase and proved that remotely sensed data improved the consistency of the predictions. Mangeas *et al.* has used non-linear artificial neural network modelling technique using remote sensing data for estimation of ecological behaviour of tuna in Indian Ocean. Using ANN models on remote sensing data like sea surface temperature, chlorophyll a concentration and sea level anomalies, they provided model of 80 per cent accuracy compared to 70 per cent and 50 per cent accuracy through best linear regression and actual fishing methodologies respectively. Similarly, Chen and Ware (1999) has demonstrated how artificial neural network can be used to improve the accuracy of the fishery stock forecasts and hence the management of the fishery resources. Mansor *et al.* (2001) used multi-criteria analysis of different oceanographic conditions (taken from remote sensing satellite data) to develop model suitable for determination of potential fishing zone in sustainable harvesting of fishery resources in South China Sea with the help of a decision support system tool called, Tropical Fish Forecasting System (TroFFS). Only problem with the application of remote sensing data for forecasting purpose is its high cost involvement, which is very difficult for most of the developing countries.

In recent years along with remote sensing data, neural network (NN) models are helping the forecasters in a great way. Neural networks are mostly data driven self-adaptive method. Neural Network Regression (NNR) is useful in explaining some non-linear relationships, where the traditional forecasting methods lack explanatory power. NNR models have been criticized for excessive training time, danger of over fitting and requirement of large number of parameters and time consumption. Another disadvantage of NN model is that it fails to explain the reasoning sometimes, even after prediction one may not know why the decision are made. However, building of parsimoniously parameterized ANN using statistical inference can solve many problems of forecasting in real life situations. Its application in agriculture and allied sector though limited or scarce but to explain and forecast the complex behavior of agricultural system NNR may play a great role in future.

Long Memory Approach

In auto-regressive integrated moving average (ARIMA) models if the integration parameter 'd' is small and greater than zero then the process exhibits a long memory, indicating to have non-negligible dependents of the observations over time; this is also known as fractionally differenced auto-regressive moving average (FARMA) model. The studies of Ray (1993) resulted that a higher order auto-regressive model are capable of forecasting the longer term well compared to fractionally integrated ARIMA (ARFIMA).

The Family of Conditional Heteroscedastic Models

Engle (1982) pioneered the auto-regressive conditional hetero-scedastic (ARCH) model. This model explained the dynamic changes in conditional variances as a deterministic function of past returns. One-step ahead forecast is readily available from this model. The additional dependency on the lagged conditional variances was allowed in a more general ARCH model called generalized auto-regressive conditional hetero-scedastic (GARCH) model. The relative forecasting performance of GARCH families of models are required to be extensively studied yet. The recent work by Galbraith and Kisinbay (2005) compared the forecast content functions from both the GARCH and (fractionally integrated) FIGARCH.

Combining Forecasts

Studies on combining forecasts, mixing or pooling as obtained from different time series methods are being studied presently. In doing so, selection of weights has become one of the major areas of studies. While simple average is widely used combining methods combining weights determined by OLS technique from the matrix of past forecast and the vector of past observations have been criticized by the possible presence of serial correlation in the combined forecast error. Zou and Yang (2004) considered combining forecast from similar models with sequentially updated weights. The future need of time series forecasting is to have extensive study on multiple time series models. Multi-variate time series forecasting models are not in considerable use because of lack of network forecasting algorithms and lack of availability of easy to use software. Routine application of multi-variate time series forecasting techniques is very difficult to estimate because of the existence of large number of parameters. There is also need for research in forecasting method based on non-linear models. The out performance of non-linear models is yet to be established emphatically. As we know that in many of the time series the Gaussian, approximation of independently and identically distributed errors with mean zero and variance σ^2 cannot be taken as granted. More over the time series data are characterized by the presence of outliers in the data that are hardly to be detected. So robust statistical procedure based on both parametric and non-parametric method is essential.

Forecast Accuracy

The accuracy of forecasting performances is measured through various constants (quantities) like mean-square-error (MSE), root- mean-squared-error (RMSE). Forecast error is defined as the difference between the observed value at time period t (Y_t) and the forecast for the period t (F_t) *i.e.* $e_t = Y_t - F_t$, and the percentage error is given by

$P_t = (e_t/Y_t) \times 100$. Sometimes different methods of forecasting are compared with the help of relative error measure and is given by $r_t = \frac{e_t}{e_t^*} = \frac{Y_t - F_t}{Y_t^* - F_t^*}$, where * denotes the difference in observed and forecasted values of the series for a standard forecasting method. Forecasting accuracy can also be measured with the help of mean absolute percentage error (MAPE), median absolute percentage error (MdAPE), mean relative absolute error (MRAE), symmetric mean relative absolute error (sMRAE) etc., each measure has certain advantages and disadvantages and most probably all are not equally efficient in every situation. For example, MSE is scale dependent hence cannot be used for comparison between two series. Similarly, MAPE has problem with the series having values closed to equals to zero. Armstrong and Collopy (1992) have recommended in-spite of having infinite variance and undefined mean the geometric mean related absolute error (GMRAE) and median related absolute error (MdRAE). The statistical properties of the measures of forecasting accuracy are yet to be worked out in an extensive way.

Example of Box–Jenkins Modelling

As has been mentioned in the subsection of time series forecasting that Box-Jenkins methods are among the important tools used in time series modelling, we take the following data of fresh water fish production of China and India to sketch the model formulation using this technique (Source: FAO data bank).

Fresh Water Fish Production of China and India During the Period 1950–2003
(Production in million tones)

Year	*1950*	*1951*	*1952*	*1953*	*1954*	*1955*	*1956*	*1957*	*1958*	*1959*
China	0.237	0.358	0.440	0.641	0.771	0.851	0.701	0.711	0.711	0.713
India	0.210	0.215	0.216	0.238	0.240	0.243	0.294	0.358	0.309	0.239

Year	*1960*	*1961*	*1962*	*1963*	*1964*	*1965*	*1966*	*1967*	*1968*	*1969*
China	0.714	0.714	0.712	0.713	0.714	0.814	0.816	0.817	0.819	0.820
India	0.282	0.277	0.330	0.390	0.460	0.507	0.478	0.537	0.622	0.693

Year	*1970*	*1971*	*1972*	*1973*	*1974*	*1975*	*1976*	*1977*	*1978*	*1979*
China	0.927	0.904	0.903	1.007	1.012	1.049	1.051	1.069	1.060	1.128
India	0.671	0.690	0.666	0.748	0.783	0.784	0.799	0.863	0.816	0.848

Contd...

Year	1980	1981	1982	1983	1984	1985	1986	1987	1988	1989
China	1.240	1.387	1.576	1.861	2.267	2.864	3.458	4.032	4.508	4.768
India	0.888	0.999	0.940	0.987	1.082	1.090	1.205	1.228	1.319	1.381

Year	1990	1991	1992	1993	1994	1995	1996	1997	1998	1999
China	5.083	5.332	5.966	7.183	8.632	10.204	11.778	13.114	14.089	15.204
India	1.574	1.655	1.682	1.883	1.931	2.116	2.229	2.340	2.435	2.677

Year	2000	2001	2002	2003
China	15.782	16.301	17.135	18.718
India	2.519	2.856	2.677	2.718

Before dealing with any time series data, it is required to check the data for existence of outlier. We used the Grubbs test for outlier check as given below:

Grubbs Test for Outlier

A good number of tests have been devised several ways to detect outliers. Grubbs' test is particularly easy to follow and is also called the ESD method (Extreme Studentized Deviate). The first step is to quantify how far the outlier is from the others. Calculate the ratio Z as the difference between the outlier and the mean divided by the SD. If Z is large, the value is far from the others. Note that we are to calculate the mean and SD from all values, including the outlier.

$$Z = \frac{|\text{mean - value}|}{\text{SD}}$$

Since 5 per cent of the values in a normal population are more than 1.96 standard deviations from the mean, our first thought might be to conclude that the outlier comes from a different population if Z is greater than 1.96. This approach only works if we know the population mean and SD. This is rarely the case in experimental science. When analyzing experimental data, we calculate the mean and SD from the data. The presence of an outlier increases the calculated SD. Since the presence of an outlier increases both the numerator (difference between the value and the mean) and denominator (SD of all values), Z does not get very large. In fact, no matter how the data are distributed, Z cannot get larger than $(N-1)/\sqrt{N}$, where N is the number of values. For example, if N=3, Z cannot be larger than 1.555 for any set of values. Grubbs and others have tabulated critical values for Z. The critical value increases with sample size, as expected. If our calculated value of Z is greater than the critical value in the table, then the P value is less than 0.05. This means that there is less than a 5 per cent chance that one would encounter an outlier so far from the others (in either direction) by chance alone, if all the data were really sampled from a single normal distribution. Note that the method only works for testing the extreme value in the sample (if in doubt, calculate Z for all values, but only calculate a P value for

Grubbs' test from the largest value of Z. Once we identify an outlier, we may choose to exclude that value from our analyses/may retain or replace it.

The result of Grubs test indicates that none of the series of our consideration has any outlier. We used graph pad software for this calculation purpose.

Next step is to check whether there exists any trend or seasonality in the time series data or not. This can be achieved by drawing graphs of autocorrelation function and partial autocorrelation functions. As a rule of thumb, the ACF and PACF are good estimate of the ACF and PACF of a stationary process for lag of about a third of the sample size. The horizontal line on the sample ACF and the PACF are the bounds $\pm\frac{1.96}{\sqrt{n}}$.

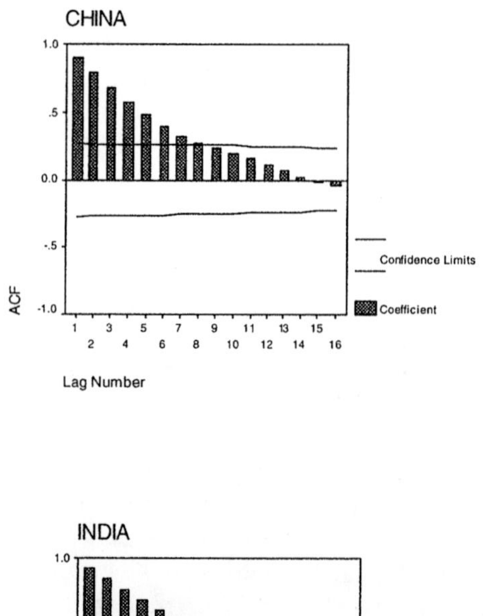

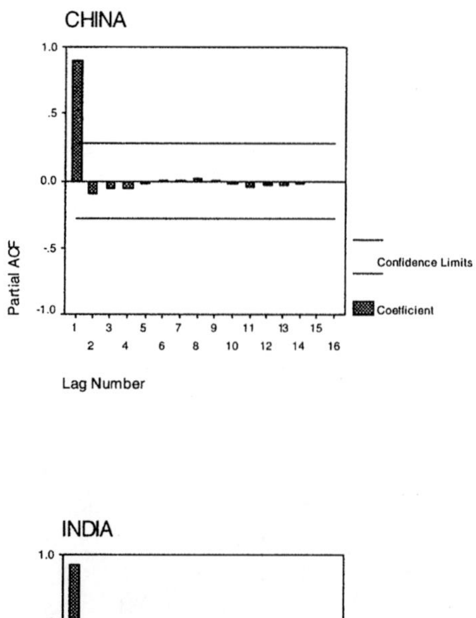

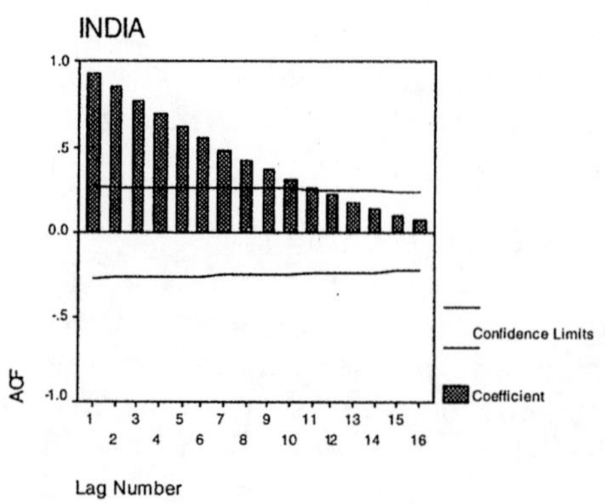

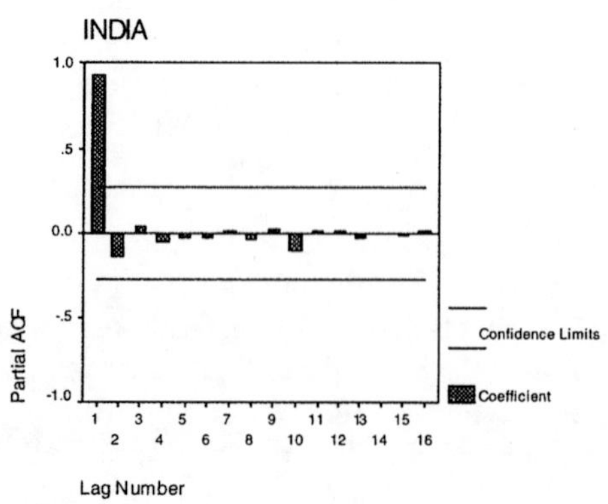

The rapid and slow decaying pattern of autocorrelation graphs for China and India respectively indicate that the production data of China is stationary, while that of India required to be made stationary before modelling.

Differencing is a technique that can be used to remove the seasonal components and the trends in a time series data. To remove the seasonal component of period 8 from the series Y_t we generate the transformed series $Y_t^* = Y_t - Y_{t-8}$. This is nothing but the differencing at lag 8. On the other hand, to eliminate linear trend from a time series we are to difference it at lag1; to eliminate quadratic trend differencing twice at lag1, and higher order polynomial trend can be eliminated analogously. We used this principle to eliminate trend and subsequently to make the series by taking first difference. Following figure shows that there exists no significant autocorrelation between the successive observations.

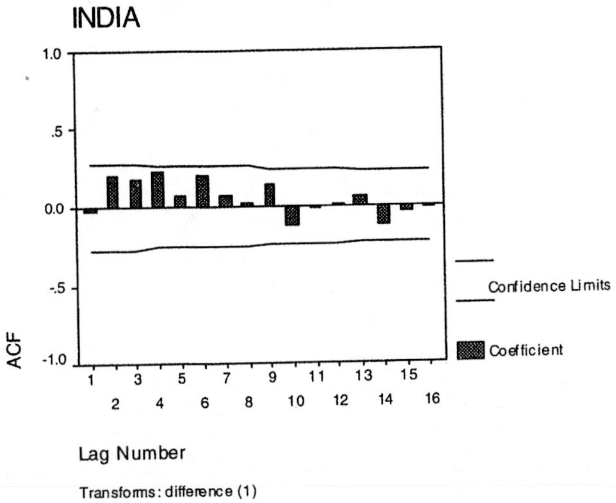

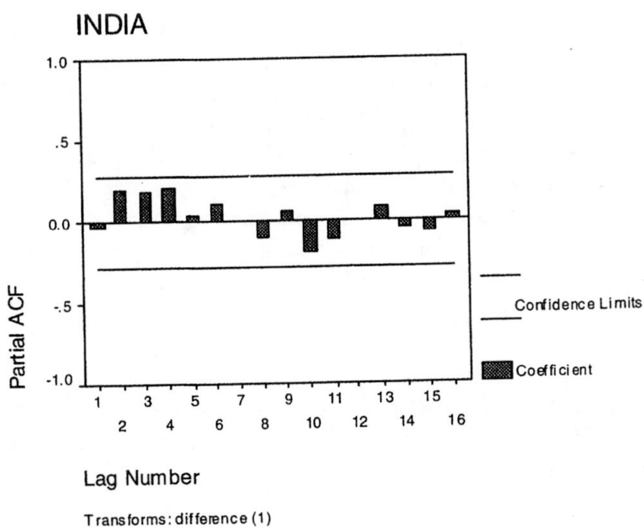

The graphs of the ACF and the PACF sometime suggest an appropriate ARIMA model for the data. As a rough guide, if the sample ACF falls between the plotted bounds $\pm\frac{1.96}{\sqrt{n}}$ for large n>q then MA (q) model is suggested, while if the sample PACF falls between the plotted bounds $\pm\frac{1.96}{\sqrt{n}}$ for large n>p, then an AR (p) model is suggested. If neither the sample neither ACF nor the PACF cuts off as mentioned above, a more refined model selection technique is required.

Let us examine the type of B-J models to be fitted with the data. The PACF graphs corresponding to both the data indicate that AR (1) model would be suitable. On the other hand ACF graphs indicate moving average MA (<9) and MA (<12) for China and India respectively are to be fitted.

Table 5.1: Model Summary

Country	*Model No*	*ARIMA Model*			*SE*	*Log likelihood*	*AIC*	*Residual*	*Remarks*
		p	*d*	*q*				*ACF PCF*	
China	1	1	0	3	0.1987	6.576	-5.151	Not white noise	Not selected
	2	1	0	6	0.1616	18.572	-23.144	White noise	Selected
India	1	1	1	0	0.0809	55.1909	-108.382	Not white noise	Not selected
	2	1	1	1	0.071	61.7914	-119.583	Not white noise	Not selected
	3	1	1	2	0.0656	65.903	-125.806	White noise	Selected

The model summary is presented in Table 5.1. Best models are selected based on Akaikes information criteria (AIC), standard error of the model, and the value of log likelihood and of course, the significance of the parameters estimated in the model. AIC, SE is kept at minimum level while log likelihood is increased. In doing so we do not sacrifices the significance of the parameters. Thus, there will be different competitive models for same set of data. Therefore, the selection of the appropriate model is being done by testing the properties of its residuals. If we are to assume that our fitted model is the true process generating the data then the observed residuals would be realized values of a white noise sequence. Precisely, the sample ACF and PACF of the observed residuals should lie with in the bounds $\pm\frac{1.96}{\sqrt{n}}$ (~ 2 SE limits roughly). These bounds are displayed on the ACF and the PACF graphs. If there are only a few or no value exceeding these lines then we may take these residuals are white noise and our model as good one, ACF and PACF of the model selected for

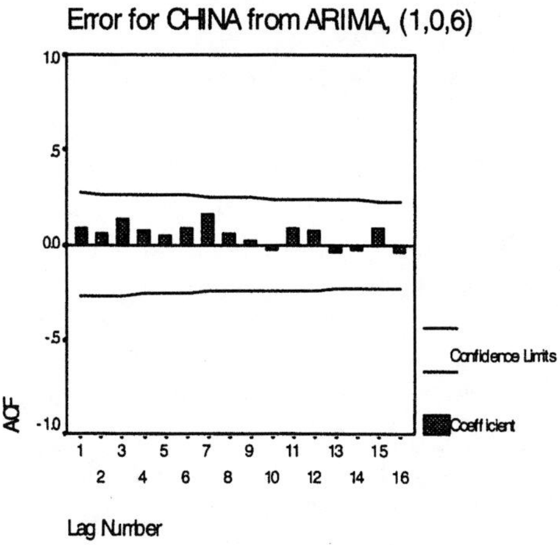

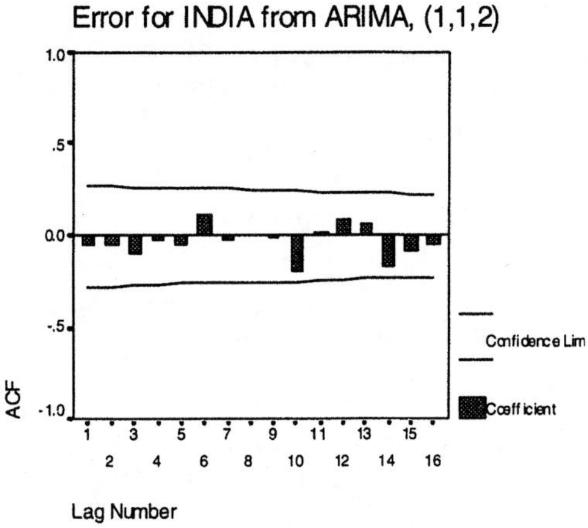

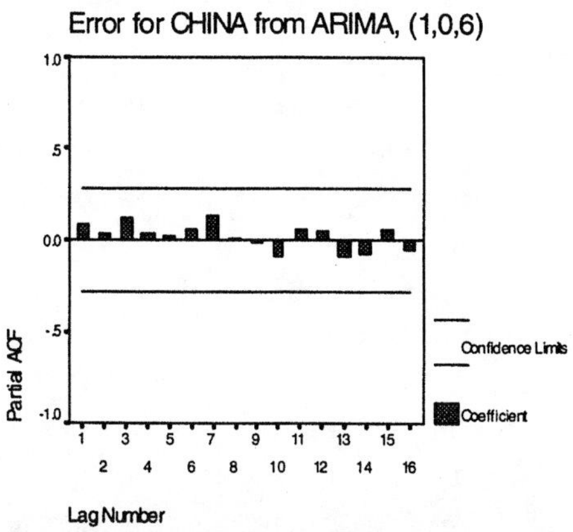

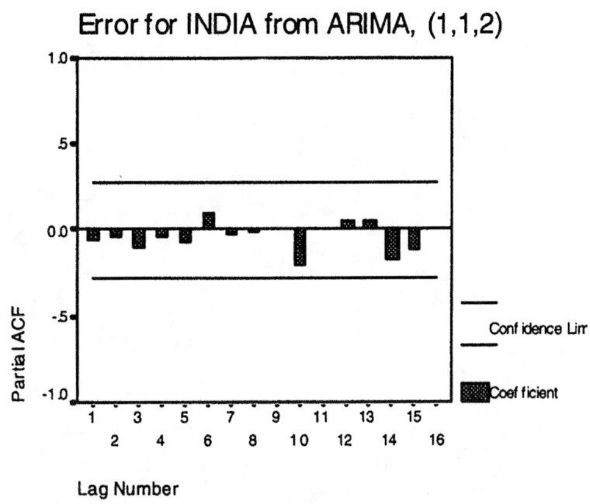

forecasting purpose of the two sets of data follow this criteria. Hence, we may take that the models are good.

Table 5.2: Observed and Predicted Production of Fresh Water Fishes in China and India (in million tones)

Year	*China*		*India*	
	Observed	*Predicted*	*Observed*	*Predicted*
2001	16.301	16.352	2.856	2.722
2002	17.135	16.695	2.677	2.792
2003	18.718	16.817	2.718	2.859
2004		16.730	2.782	2.923
2005		16.614	2.782	2.984
2006		16.335		3.042
2007		16.142		3.098
2008		15.952		3.151
2009		15.763		3.201

This is again proved with the help of forecasted values close to the observed values for the model validation period, that is 2001 to 2003. Our models have more accurately predicted the production of India than China. Using these models, we can forecast the short-term production of fresh water fish production in India and China for years to come. We used the statistical software SPSS for modelling purpose.

This is an example, in brief, how we can go for time series modelling with the help of Box-Jenkins's technique but certainly this is not a unique one. As has been mentioned we are to go through repeated trials to get suitable model, surely one can have better model than these two.

References

Armstrong, J. S. and Collopy, F., 1992. Error measures for generalizing about forecasting methods: Empirical comparisons. *International Journal of Forecasting,* 8:69-80.

Barron, 1993. Universal approximation bounds for superpositions of sigmoid function. IEEE Transaction on Information Theory 39.

Bolger, F. and Wright, G., 1994. Assessing the quality of expert judgment: issues and analysis. *Decision Support System,* 11:1-24.

Box, G. E. P. and Jenkins, G. M., 1976. Time series analysis: Forecasting and control. San Francisco: Holden Day.

Bradley, M. D. and Jansen, D. W., 2004. Forecasting with a non-linear dynamic model of stock returns and industrial production. *International Journal of Forecasting,* 20: 321-342.

Brazil, C., F. West, and Baker, T., 2006. 2007 Togiak herring forecast. *A release by Alaska department of fish and game division of commercial fisheries on 11.9.06.*

Brugere, C. and Ridler, N., 2004. Global aquaculture outlook in the next decades: an analysis of national aquaculture production forecasts to 2030. *FAO Fisheries Circular No 1001,* Rome.

Chand Ramesh, 2004. Agricultural Growth during the Reforms and Liberalization: Issues and Concerns, *Policy Brief* (20), *NCAP, ICAR,* New Delhi, India.

Chen, D. G. and Ware, D. M., 1999. A neural network model for forecasting fish stock recruitment. *Can. J. Fish. Aquat. Sci.,* 56(12):2385-2396.

Clements, M. P. and Smith, J., 1997. The performance of alternative methods for SETAR models. *International Journal of Forecasting,* 13:463-475.

De Silva, S., 2001. A global perspective of aquaculture in the new millennium. In Subasinghe R P, Bueno P, Phillips M J, Hough C, McGladdery S E and Arthur J R (Eds.). Aquaculture in the next Millennium. *Technical Proceedings of the Conference on aquaculture in the Third Millennium,* Bangkok, Thailand, 20-25 February 2000, NACA, Bangkok and FAO, Rome, pp 431-459.

Delgado, C. L., Wada, N., Rosegrant, M. W., Meijer, S. and Ahmed, M., 2003. Fish to 2020. Supply and Demand in Changing Market. *International Food Policy Research Institute, Washington DC and World Fish Center,* Penang.

Doraiswamy, P. C., Moulin, S., Cook, P. W. and Stern, A., 2003. Crop yield assessment from remote sensing. *PE and RS photogramatic engineering and remote sensing*, 69(6):665-674.

Engle, R. F., 1982. Autoregressive conditional heteroscedasticity with estimates of the variance of the United Kingdom inflation. *Econometrica*, 50: 987-1008.

Galbraith, J. W. and Kisinbay, T., 2005. Content horizons for conditional variance forecasts. *International Journal of Forecasting*, 21:249-260.

Gardner, Jr E. S., 1985. Exponential smoothing: The state of the art. *Journal of Forecasting*, 4:1-38.

Gotsch, N. and Rieder, P., 1990. Forecasting future developments in crop protection, *Crop protection*, 9(2): 83-89.

Harvill, J. L. and Ray, B. K., 2005. A note on multi-step forecasting with functional coefficient autoregressive models. *International Journal of Forecasting*, 21: 717-727.

Howling, G. G., Harrington, R., Clark, S. J. and Bale, J. S., 1993. The use of multiple regression via principal components in forecasting early season aphid (Homopter: Aphididae) flight. *Bulletin of Entomological Research*, 83(3):377-381.

http://www.graphpad.com/quickcalcs/Grubbs1.cfm

Kim, J. H., 2003. Forecasting autoregressive time series with bias corrected parameter estimators. *International Journal of Forecasting*, 19:493-502.

Lawrence, M., Edmundson, R. and O'Conor, M., 1985. An examination of the accuracy of the judgemental extrapolation of time series. *International Journal of Forecasting*, 1:25-35.

Mansor, S., Tan, C. K., Ibrahim, H. M. and Md. Shariff, A. R., 2001. Satellite Fish Forecasting In South China Sea. Paper presented at the 22nd *Asian Conference on Remote Sensing*, 5-9 November 2001, Singapore.

Newbold, P., Agiakloglou, C. and Miller, J., 1994. Adventure with ARIMA software. *International Journal of Forecasting*, 10:573-581.

Pal, S. and Sahu, P. K. 2004. Modelling pest incidence–a nonparametric approach, *Abstracted in the 24th ISF*, held in Sydney, 2004.

Parsons, D. G. and Colbourne, E. B., 2000. Forecasting Fishery Performance for Northern Shrimp (*Pandalus borealis*) on the Labrador Shelf, (NAFO Divisions 2HJ). *J Northw.Atl.Fish.Sci*, 27:11-20.

Ray, B. K., 1993. Long range forecasting of IBM product revenues using a seasonal fractionally differenced ARMA model. *International Journal of Forecasting*, 9:255-269.

Roger, Z., 2005. Forecast for the 2005 brown shrimp season in the western Gulf of Mexico, from the Mississipi River to the US.–Mexico Border. *A release by the director, Galveston Laboratory*. http://galveston.ssp.nmfs.gov/galv.

Sarantis, N., 2001. Nonlinearities, cyclical behaviour and predictability in stock markets: International evidence. *International Journal of Forecasting*, 17:459-482.

Sender, N. R. and Manrodt, K. B., 2003. The efficiency of using judgmental versus quantitative forecasting methods in practice. *Omega*, 31:511-522.

Siddiq, E. A., 2002. Rice–Exploring means to adopt G M Rice. In *The Hindu Survey of Indian Agriculture*, pp 47-52.

Snyder, R. D., 1985. Recursive estimation of dynamic linear statistical models. *Journal of Royal Statistical Society* (*B*) 47:272-276.

Wang, Z. and Bessler, D. A., 2004. Forecasting performance of multivariate time series models with a full and reduced rank: an empirical examination. *International Journal of Forecasting*, 20:683-695.

Wijkstrom, U. N., 2003. Short and long-term prospects for consumption of fish. *Veterinary Research Communications*, 27 Suppl.1:461-468.

WMO, 1992.International Meteorological Vocabulary. No18, Geneva, Switzerland.

Ye, Y., 1999. Historical consumption and future demand for fish and fishery products: Exploratory calculations for the years 2015-2030. *FAO Fisheries Circular No 946*, Rome, FAO.

Yule, G. U., 1927. On the method of investigating periodicities in disturbed series with special reference to Wolfer's support numbers. *Philosophical transactions of the Royal Society of London, Series A* 226:267-298.

Zou, H. and Yang, Y., 2004. Combining time series models for forecasting. *International Journal of Forecasting*, 20:69-84.

Modelling, Forecasting, Artificial Neural Network and Expert System in Fisheries *Pages 67–86*
Editors: **Ajit Kumar Roy & Niranjan Sarangi**
Published by: **DAYA PUBLISHING HOUSE (2009)**

Chapter 6

Modelling in Aquaculture

A.K. Roy

Introduction

During the past decade aquaculture has been involved in developing breeding and culture technologies for growing and production of various indigenous and exotic species of carp and prawn. In the process a good knowledge and understanding has been achieved regarding how growth is affected by nutrition, feed rate, temperature, humidity, season of the year, species composition, ratio, aeration, etc. It may be stated that a good understanding exists between state variable, central variable and parameters. Therefore, the modelling, identification and control of many aquaculture process and facilities is relatively easier and useful. It is well known that aquaculture is a mixture of species and facilities and environment. Each species has unique requirements for habitat, environment and nutrition. The reproduction, physiology and pathology may vary between different cultured species. The facilities ranges from backyard ditch to tank to ponds, from indoor to outdoor, from freshwater to brackish water to saltwater, from confined to ocean, from monoculture to polyculture. The processes that take place in these facilities are in the most part are similar yet some are diverse *e.g.*, feeding peletted diet to prawn, plankton to nursery reared one, spawning fry fish vs. seeding shellfish. Some aquaculture system is intensive, some extensive and some semi intensive. On the other hand a wide variety of aquaculture systems exists like ponds, cages, tanks, enclosures, rice fields etc. These systems are difficult to study directly because of their size and complexity. A diversity of biota comprising of fish, phyto and zooplankton, benthos, macrophytes, insects, bacteria etc. coexist and interact in an aquatic environment characterized by many physical, chemical factors like temperature, dissolved gases, alkalinity, pH, etc that can vary not only in three dimensional space but also in time with daily and seasonal components (Boyd, 1979). Because of these reasons probably aquaculture pond is generally considered as a 'black box'. Therefore 'black box' research approach of data exploration or knowledge discovery is generally taken in order to understand the aquaculture system/process.

Concept of Modelling

Modelling is an interdisciplinary approach to problem solving characterized by the construction, testing and use of models (Riggs 1963; Spain 1982; Grant 1986). It is a tool to specify, describe, organize and communicate knowledge about complex phenomena in precise, clear and concise form.

An important principle of modelling is that a mathematical equation serves as a model of a biological process (Spain 1982). A biological process may be described graphically. The graph expresses how a biological system performs under a given set of conditions. A graph can also describe the behavior of a mathematical equation relating an independent to a dependent variable. Thus, it is possible to find a mathematical equation that represents the biological system. The process of finding an appropriate model involves an understanding of the functional relationships in the system. After examination of experimental data, a simple equation relating a dependent variable Y to an independent variable X can be formulated. The parameters for the equation (the coefficients and exponents) can be estimated from experimental data, after which the model must be validated by comparing data generated by the model with data from the real system. Finally, the model can be sued to make predictions about the behaviour of the real system. Modelling is an important step in accelerating the use of more quantitative and precise methods in aquaculture research. Mathematics provides models with a clear and precise language for describing and examining a system (Spain 1982; Karplus 1983).

Types of Models

Conceptual and Mathematical Models

Using the broad definition of a model, it is evident that essentially all science deals with the fornulation, examination and improvement of conceptual models of our universe (Spain 1982). The conceptual model is a picture of the real system based on available information and past experience. Conceptual models by themselves are generally lacking in precision and rigor (Spain 1982). A conceptual model needs to be translated into a form that is subject to precise description, evaluation and validation. This form of model is called a *mathematical model*. Such models may be as simple as a single equation relating one variable to another or may involve the interaction of many equations having several mutually dependent variables (Spain 1982).

Empirical and Theoretical Models

Empirical (also called statistical or descriptive) models are constructed by fitting one of several possible forms of equations to experimental data using statistical techniques. Empirical models merely describe the relationship between two or more variables as it was actually observed (Riggs 1963). Models do not depend upon a theoretical understanding of the system (Spain 1982) and imply nothing about the underlying reason for the relationship. The system is treated as a "black box" : inputs and outputs are known, but the underlying mechanisms are not known (Piedrahita 1988). A particular data set can be described by an infinite number of empirical models. When no guidance is available from theory, the simplest equation which adequately fits the data should be chosen. Empirical models can contribute to biological theory since a model that summarizes an observed relationship may serve as the point of departure for the elaboration of a theory.

Theoretical (also called *mechanistic* or *explanatory*) models are based upon a theory or hypothesis about the nature of a system (Riggs 1963; Spain 1982). Whereas an empirical model can describe a particular data set faithfully enough. A well-founded theoretical model may describe all such sets of data and in addition explain why the observed relationships exist. Failure of a theoretical model to fit a wide range of experimental values can be turned to advantage if it suggests how the underlying theory can be improved. Theoretical models play an essential part in the formulation and solution of biological problems that are too complex to solve nonmathematically. Because of their generality and power to aid in the development of basic knowledge, theoretical models have an elegance and intellectual appeal that is lacking in empirical models.

Deterministic and Stochastic Models

A deterministic model uses fixed values for model parameters and produces the same results given the same set of input variables (Spain 1982; Starfield and Bleloch 1986). In such models, the stage of a given variable at any time is determined entirely by previous states of that variable and the other variables upon which it depends.

A stochastic (also called probabilistic) model uses random values for model parameters. Repeated calculations with the same input variables will therefore yield different output variables with every calculation. Probabilistic models are generally more complex than deterministic models because of the mathematics involved in formulating the random process. A biological process may appear to be random because it is impossible to predict its behavior from the information currently available. As more is known about a system, a process previously thought of as random may turn out to be determined by biological factors.

Static and Dynamic Models

A static (also called time-independent) model does not change with time. A dynamic (or time-dependent) model changes as a function of time. Dynamic models can be discrete or continuous. In continuous dynamic models, time flows.

Discrete and Continuous Model

A discrete model is simpler in many ways and corresponds to the way data are usually collected in batches at regular intervals. Continuous models can be approximated, choosing a time interval small enough to capture the behavior of a system across time.

Determination of the Variables of Aquaculture Models

To determine what to include or exclude from an aquaculture model, it is helpful to look at things from the position of the cultured organism. Fish depend upon their environment for the supply of resources (food, oxygen), for removal of metabolic wastes (ammonia, carbon dioxide, feces, urine) and for maintenance of conditions suitable for growth, survival and reproduction (temperature, pH, salinity, etc.).

Knowledge of Types of Variables

It is useful to make the distinction between *state variables, rate variables, driving variables and parameters. State variables* are the quantities in the model that determine the state of the model at any point in time. Rate variables represent the flow of materials from or to state variables. Driving variables (or auxiliary variables) and parameters are not affected by the system. Parameters are given constants and driving variables can be changed by the modeler to simulate certain conditions (*e.g.*, a feeding regime or a climate). Classifying environmental factors according to their action on any biological function and not merely by type grouping is useful in the understanding of biological systems (Brett 1979).

Controlling factors govern the rates of biological reactions and operate at all nonlethal levels of the factor concerned. A distinction can be made between external and internal controlling factors. Examples are temperature, which sets the pace of all metabolic processes in fishes; and body size, which scales the metabolic rate to the size of the fish, respectively.

Limiting factors restrict the supply of food and oxygen to the fish or the removal of its metabolic wastes. They become operational at specific levels. Unionized ammonia and carbon dioxide may restrict the removal of these metabolic wastes.

Models of an aquaculture enterprise having together models of three major subsystem namely (*i*) biological models (*ii*) physical models and (*iii*) economic model.

Biological Model

Biological models are concerned with the responses (growth, survival, reproduction) and interactions (eating, breathing, excretion) of the cultured fish and its environment.

- ☆ *Choice of species:* the biological requirements of the species have to be matched to its proposed environment as defined by site selection and facility design.
- ☆ *Seed or fingerling production:* this phase may include reproduction in captivity and hatchery operations to raise the larvae/fry.
- ☆ *Growout to market size:* the aquaculture production phase most commonly modeled is the growout from fry or fingerling to market-size fish.

Physical Model

The physical model describes mathematically how the physical process and facility relate to the aquatic environment. They simulate the relationship between the control variables and state variables.

Pond Ecosystem Models

According to Cuenco, 1984, pond ecosystem models attempt to incorporate the components and processes relevant to the pond environment. These include the biotic components (cultured organisms, phytoplankton, zooplankton, benthic organisms, bacteria-detritus complex), the abiotic components (water characteristics, such as pH, alkalinity, carbon dioxide, temperature, dissolved oxygen, ammonia-ammonium, nitrite, nitrate, phosphorous, organic matter; soil type; climatic factors such as solar radiation, wind, evaporation and precipitation) and management factors. The core of a pond model is usually some type of growth model. Linked to this are the environmental factors that are considered important or of interest for the particular problem under study, and the management factors (especially inputs like feeds and fertilizers). The environmental and management factors that are usually incorporated in pond models are described here briefly (Cuenco, 1984).

Dissolved Oxygen (*DO*) is required by the fish for its metabolism. Fish growth rate increases with rising oxygen concentration up to a maximum where oxygen is not limiting anymore and growth rate is independent of the oxygen concentration.

Unionized Ammonia (*NH_3*) is a by-product of metabolism and can be regarded as a limiting factor restricting the removal of a metabolite and consequently decreasing growth (Cuenco 1982). Unionized ammonia is in equilibrium with the ammonium ion (NH_4+). High temperatures and high pH cause the equilibrium to shift towards NH_3, which is very toxic to fish.

The *pH* of the water in freshwater tropical fishponds shows a diurnal rhythm, which is caused by the production and consumption of carbon dioxide. During the daytime, carbon dioxide is taken up by the phytoplankton causing an increase in pH. At night, photosynthesis stops while respiration continues to produce carbon dioxide, causing the pH to decrease. The pH level affects the ammonia equilibrium.

Growth, like other physiological processes, is regulated by body temperature in fish which is equal to the ambient water temperature. Relative growth rate increases with rising temperature, reaches a peak at the optimum temperature for growth and falls steeply above the optimum temperature (Brett and Groves 1979).

Of the *management factors*, pond fertilization (Swingle and Smith 1939; Hepher 1962) and *feeding* (Shell 1968; Swingle 1968) are traditional method of increasing fish production in ponds. The distinction between fertilization and feeding is arbitrary because uneaten feeds add fertility to ponds (Boyd 1979) and some organic fertilizers may be eaten directly by the fish (Spataru 1976). Frequent applications of small manure doses give higher fish yields than applying large amounts a few times during the growing season. Other management factors are species ratio, species composition, stocking density and harvesting technique. The ideal stocking density is the maximum density that results in production of desirable-sized fish in a reasonable period of time (Swingle 1968). The maximum number of fish that can be stocked depends on the maximum biomass the pond can support (*i.e.*, the carrying capacity) and the average size of the fish. As fish size increases during the culture period, the initial biomass may be far below the carrying capacity of the pond. With a model, it is possible to investigate the possibility of stocking more fish of different sizes with regular harvesting during the culture period to regulate the biomass.

Constraints to Modelling Aquaculture Systems

Modelling is trying to characterize the system given the excitation and response (Karplus 1983). Modelling is therefore a synthesis problem with no unique solution and the number of mechanisms that are incorporated in the model determines the position of the model in the continuum from empirical "black box" models to theoretical "white box" models (Karplus 1983).

Management Applications

The Need to Manage Nature : Models may be developed to facilitate the day-to-day management of aquaculture operations, *e.g.*, determining stocking and feeding rates, predicting dissolved oxygen levels and examining the effects of different management strategies (Piedrahita 1988). Decision-making involves an implicit, if not explicit use of models, since the decision makers invariably have a causal relationship in mind when they make a decision (Karplus 1983). On the other hand, models cannot replace decision makers and a model should never be used as a substitute for good judgment. Very often some other factors have to be considered along with the model results before committing to a course of action (Phillips *et al.*, 1976). The subjective decisions and assumptions made during construction of the model are often not met fully in the particular problem to which the model is applied. The model is merely a tool: the modeler has to make the decisions. Eventually, aquaculture models should be developed to the point where they become important tools in designing and managing aquaculture systems.

Economic Model

The success of aquaculture is measured ultimately by profits. Management and operational decision must be founded upon their economic impact. The purpose of the economic model is to incorporate costs of physical inputs and value of outputs produced. The most important thing is that operational manipulation of control variables must reflect the subsequent impact on costs and revenues.

Validation

Validation is the process of testing of how much confidence can be placed on the model results that are consistent with reality and meaningful when properly interpreted.

For actual statistical modelling, many methods exist (Figure 6.1). Aquaculture researchers have used the following methods on their own data: multiple regression analysis (Hopkins and Cruz 1982; Pauly and Hopkins 1983; Prein 1985; Gonzat *et al.*, 1987;), path analysis (Prein 1985), principal component analysis (Milstein *et al.*, 1985a, 1985b) and canonical correlation analysis (Milstein *et al.*, 1988). Many textbooks on statistics also cover some parts of multivariate statistical techniques (Yamane, 1973; Morrison 1976; Snedecor and Cochran 1980; Hair *et al.*, 1987).

Figure 6.1: Methods for Statistics Modelling

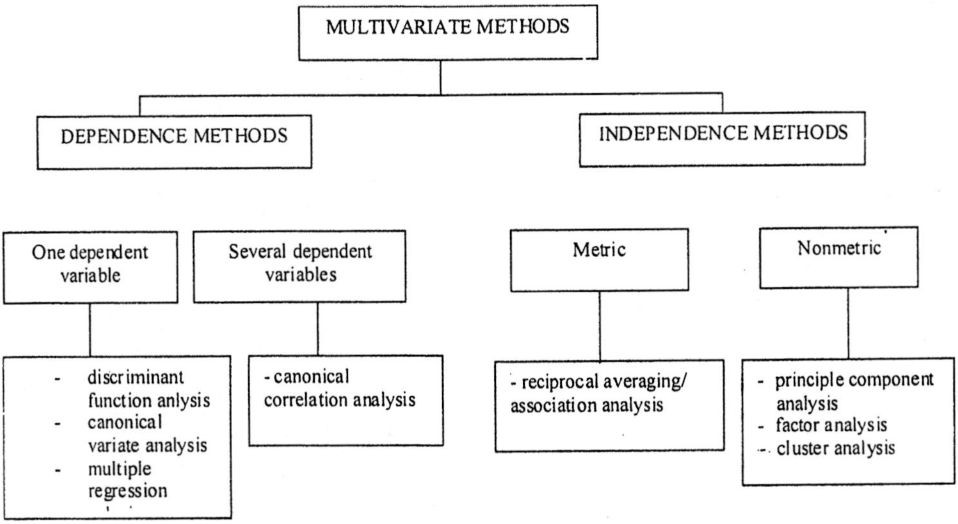

(*Source:* Adapted from Jeffers 1982 and Hair *et al.*, 1987)

Multivariate Exploratory Techniques designed specifically to identify patterns in multivariate data sets are cluster analysis, factor analysis, discriminant function analysis, multidimensional scaling, linear analysis, canonical correlation, stepwise linear and non-linear analysis, regression analysis, correspondence analysis, etc.

Exploratory Methods

Exploratory methods are mostly applied to datasets with numerous variables collected to answer specific questions. Here methods are used to reduce the large amount of individual information contained in distinct variables and objects into a few components representing the major relationship in the data. Factor analysis is used to reduce the number of variables in a dataset to few central variables. Cluster analysis combines traits of objects and not variables to describe groups. Multidimensional scaling only searches for global similarities between objects and aims to depict the measured similarities in a space of least possible dimension.

The sequence of steps in this above mentioned example is described in more detail in a number of modelling studies. For examples of theoretical aquaculture system models see: Cuenco *et al.* (1985a, 1985b, 1985c); Piedrahita (1988); Machiels and Henken (1986, 1987); Machiels and van Dam (1987); Machiels (1987). More general models of freshwater ecosystems are described in *e.g.* LeCren and Lowe Mc Connell (1980). Latest studies include projection of fish production, consumption, and supply and demand. (Delagado *et al.*, 2003; Charn *et al.*, 2002; Garcia and Newton, 1997, and Ye and Beddington, 1996). Some reports are available on the modelling in aquaculture in India (Roy *et al.*, 1998, 2000, 2006; Bhar *et al.*, 2002).

Empherical Example of Developing a Biophysical Model Using Multivariate Method on Fish Growth Data of Aquaculture Experiment

In this section steps in biophysical modelling with the help of an illustrated example have been described. The dataset taken for identifying and quantifying the influence of various biotic and abiotic factors on the growth of *L. rohita* cultured in sewage fed fish culture ponds of Wastewater Aquaculture Division of CIFA, Rahara, 24 Parganas (N), West Bengal. The data consists of thirty-six sets of monthly weight increment expressed as specific growth rate (SGR). Sampling data along with corresponding representative values of biotic and abiotic factors are also taken from a sewage fed aquaculture ponds. A sample fifteen (15) sets of dependent and independent variables are presented in Table 6.1 for getting an idea about the nature of data.

Multiple Linear Regression

In order to study the complex aquaculture problems covering fish growth, pond ecology and environmental parameters exploratory data analysis is done using multiple linear regression models which can be expressed as $\hat{Y}=a+b_1X_1+b_2X_2+\dots+b_{11}X_{11}$

Where $\hat{Y}$ is Specific Growth Rate (per cent per day) of *Labeo rohita* and b_1, b_2, ... and b_{11} are the regression coefficients of the independent variables X_1, X_2, ..., and X_{11} respectively. Again where X_1 = DO (ppm), X_2 = BOD (ppm), X_3 = pH, X_4 = NH_4-N (ppm), X_5 = P_2O_5 (ppm), X_6 = Alkalinity (ppm), X_7 = Chloride (ppm), X_8 = Free CO_2 (ppm), X_9 = Temperature (°C), X_{10} = Plankton (no./l: log transformed value), X_{11} = Initial stocking weight of Rohu

Multiple regression analysis following forward selection, backward elimination and stepwise regression were done using SPSS software package and using standard methods (Draper and Smith, 1981). The model assumes that there is a normal distribution of the dependent variable for every combination of the values of independent variables in the model. For example if fish weight is dependent variable and plankton concentration and dissolved oxygen (DO) are the independent variables, it is assumed that for every combination of plankton and DO, there is a normal distribution of fish weight and though the means of these distributions may be different but all have same variance.

At first descriptive statistics of the biotic and abiotic factors used in the present study are presented in Table 6.2.

Correlation Matrix

Secondly for developing a model and for understanding the complex aquaculture process with all independent variables is to calculate a correlation matrix for all variables. Table 6.3 shows the matrix of correlation between the dependent variable (SGR) and each independent variables as well as the correlation between the independent variables. A perusal of Table 6.3 shows that significant correlation exist between SGR and each of DO, NH_4-N, INTWT and LnPL. Significant correlation between the independent variables between BOD and Alkalinity; DO and Initial wt;

Table 6.1: A Sample of 15 Sets of Specific Growth Rate of *Labeo rohita* and Average Values of Various Biotic and Abiotic Factors

BOD (ppm)	DO (ppm)	pH	Av.P (ppm)	NH_4N (ppm)	Alkalini (ppm)	Chloride (ppm)	INITWT (g)	SGR (%/day)	LnPL (No./L)	$FreeCO_2$ (ppm)	Temp. (0°C)
8.60	12.40	8.00	0.13	0.40	160.00	100.00	22.00	2.5161	6.7979	1.00	30.00
9.60	13.60	7.40	0.54	0.40	200.00	100.00	18.50	3.2468	7.3251	1.50	25.20
14.00	6.80	8.40	1.70	6.00	340.00	100.00	22.00	4.9559	5.2983	4.00	18.60
9.00	8.00	8.00	0.10	0.12	240.00	115.00	46.80	2.6327	5.4596	4.00	18.00
8.00	8.00	7.80	0.10	0.10	205.00	110.00	49.00	2.4242	6.4489	2.00	24.00
12.00	7.20	8.40	0.94	0.10	360.00	100.00	97.30	1.5044	5.8777	2.00	26.40
7.80	7.00	7.80	0.10	0.09	240.00	110.00	103.10	0.9718	4.4188	10.00	30.00
7.60	7.00	7.70	0.15	0.20	200.00	100.00	101.40	0.5835	3.5264	6.00	25.20
28.00	1.20	8.40	0.95	0.18	300.00	110.00	152.80	0.3516	3.9890	6.20	18.80
10.00	5.20	8.20	0.42	1.50	260.00	110.00	209.20	0.5952	5.9890	4.00	29.00
10.40	8.00	7.00	0.10	0.90	220.00	110.00	91.00	0.3738	6.3630	2.00	30.00
6.00	6.80	8.00	0.05	0.08	226.00	130.00	43.40	0.6090	3.2581	2.00	25.20
8.80	6.00	7.80	0.15	2.00	220.00	120.00	201.10	1.5567	5.9506	10.00	19.00
12.00	9.20	7.70	0.13	1.20	150.00	100.00	101.80	0.7738	3.1781	4.00	17.80
8.00	6.80	8.00	0.03	0.00	220.00	100.00	52.10	1.3537	4.9053	2.00	23.80

and Av.P. and NH_4-N can substantially affect the results of a multiple regression analysis.

Table 6.2: Descriptive Statistics of Biotic and Abiotic Factors Used in the Present Analysis of *L. rohita* Culture Experiment

Variable	*Mean*	*Std Dev*	*Range*	*Minimum*	*Maximum*
AV.P (ppm)	35	0.46	2.08	0.03	2.10
NH4_N(ppm)	0.99	1.25	6.00	0.00	6.00
SGR (per cent per day)	1.412	1.363	7.42	0.29	7.71
Free CO_2 (ppm)	3.34	2.43	9.00	1.00	10.00
LNPL (No./l)	5.35	1.89	7.49	2.3026	9.7965
DO (ppm)	7.44	2.67	13.20	1.20	14.40
pH	7.95	0.37	1.50	7.00	8.50
BOD (ppm)	10.17	4.39	23.20	4.80	28.00
Temp. (°C)	23.76	4.36	12.40	17.80	30.20
Chloride (ppm)	109.03	11.20	50.00	100.00	150.00
INITWT (g)	207.42	160.65	486.90	15.00	501.90
Alkalnity (pip)	230.56	66.182	64.00	126.00	390.00

SGR: Specific growth rate; LnPL: Log transformed value of Plankton; INITWT: Initial weight of Rohu; TEMP: Temperature

Determining Important Variables

In multiple regressions, one may want to assign relative importance to each independent variable. For example one might want to know whether plankton concentration is more important than DO level in a pond in predicting SGR of fish. In this situation two approaches are there depending on which of the following questions are asked. (*i*) How important are DO and Plankton when each one is used alone to predict SGR, (ii) How important are DO and Plankton when they are used to predict SGR along with other independent variables in the regression equation. The first question can be answered by examining the correlation coefficient between SGR and each of Plankton and DO. It is apparent from the Table 6.3 that strong correlation exists between SGR and Av.P (0.655), SGR and DO (0.604) and SGR and NH_4-N (0.464) and SGR and plankton (0.405) and negative significant relationship between SGR and initial weight (–0.520). It is well known that the larger the absolute value of the correlation co-efficient, the stronger the linear association. Therefore, one would assign Av.P more importance as a predictor variable of SGR followed by DO, initial wt, NH_4-N and plankton. When the independent variables are correlated among themselves, the unique contribution of each is difficult to assess and second question is considerably complicated.

Beta Co-efficient

It is inappropriate to interpret the B's (regression coefficients) as indicators of relative importance of variables. Actual magnitude of co-efficient depends on the

Table 6.3: Correlation Matrix of the Biotic and Abiotic Parameters of Sewage Fed Eco-system

	SGR	*BOD*	*DO*	*pH*	*Av.P*	*NH_4_N*	*Alkalinity*	*Chloride*	*INITWT*	*LnPL*	*Temp.*	*Free CO_2*
SGR	1.000											
BOD	–0.028	1.000										
DO	0.604	–0.316	1.000									
PH	0.324	0.311	–0.078	1.000								
AV.P	0.655	0.340	0.179	0.386	1.000							
NH_4_N	0.464	0.182	0.193	0.087	0.660	1.000						
ALKA	0.152	0.520	–0.248	0.516	0.519	0.239	1.000					
CHLO	–0.166	0.070	–.0204	0.023	–0.265	–0.151	–0.096	1.000				
INITWT	–0.520	0.272	–0.492	–0.177	–0.166	0.131	–0.010	0.270	1.000			
LnPL	0.405	0.053	0.321	0.076	0.256	0.209	0.069	0.281	0.111	1.000		
TEMP	0.007	–0.272	0.298	–0.100	0.029	–0.193	0.033	–0.150	–0.099	0.068	1.000	
$FRCO_2$	0.183	0.169	–0.312	–0.105	–0.128	0.094	–0.080	0.133	0.552	–0.178	–0.112	1.000

units in which the variables are measured. If all independent variables are measured in the same units their co-efficient are directly comparable. Regression co-efficient of different units are comparable when all variables are expressed in standardized (Z score). They are also affected by the correlation of independent variables and do not reflect the importance of the various independent variables.

Part and Partial Co-efficients

Another way of assessing the relative importance of independent variables is to consider the increase in R^2 when a variable is entered into an equation that already contains other independent variables. This increase is R^2 change = $R^2 - R_{(i)}^2$ where $R_{(i)}^2$ is the square of the multiple correlation co-efficient when all independent variables except i[th] are in the equation. A large change in R^2 indicates that a variable provides unique information about the dependent variable that is not available from the other independent variables in the equation. The signed square root of the increase is called the partial correlation co-efficient. It is the correlation co-efficient between Y and X_i when linear effects of the other independent variable have been removed from X_i. If all the independent variables are uncorrelated, the change in R^2, when a variable is entered into the equation is simply the square of the correlation co-efficient between that variable and the dependent variable. The values of R^2 change presented in Table 6.4 and 6.5 indicates that with the addition of each of the variable to an equation, how much R^2 change occurs. It may be noted that it does not indicate what proportion of unexplained variation this increase constitutes. From Table 6.5 it can be seen that R^2 change in case of temperature is maximum followed by BOD, pH, and alkalinity.

Table 6.4: Summary Statistics of Backward Elimination

Method: Enter

Step	MultR	Rsq.	AdjRsq.	F(Eqn)	SigF	RsqCh	FCh	SigCh		Variable	Betaln	Correl
1									In:	Free CO_2	−0.1830	−0.1830
2									In:	Alkalinity	0.1380	0.1517
3									In:	Temp.	−0.0169	0.0071
4									In:	LNPL	0.3794	0.4047
5									In:	Chloride	−02948	−01656
6									In:	NH4_N	0.3725	0.4636
7									In:	pH	0.3565	0.3243
8									In:	BOD	−0.1454	−0.0284
9									In:	INITWT	−0.7320	−0.5201
10									In:	DO	0.2069	0.6042
11	0.9195	0.8455	0.7747	11.942	0.000	0.8455	11.942	0.000	In:	AV.P	0.4351	0.6553

Building a Model

Before considering several formal procedures for model building, it is necessary to examine some of the consequences of adding and deleting variables from regression

Table 6.5: Summary Statistics of Backward Elimination

Method: *Backward* *Criterion* *POUT.1000*

BOD *DO* *pH* *AV.P* *NH4_N* *ALKALI* *CHLO* *INTWT* *LnPL* *TEMP* *FREECO*$_2$

Step	*MultR*	*Rsq*	*AdjRsq*	*F(Eqn)*	*SigF*	*RsqCh*	*FCh*	*SigCh*	*Variable*	*Betaln*	*Correl*
12	0.9195	0.8454	0.7836	13.673	0.000	–.0001	0.016	0.900	Out:	CHLORIDE	–0.1656
13	0.9188	0.8441	0.7902	15.644	0.000	–0.0013	0.210	0.651	Out:	ALKALINI	0.1517
14	0.9172	0.8413	0.7943	17.898	0.000	–0.0028	0.462	0.503	Out:	NH4_N	0.4636
15	0.9125	0.8327	0.7909	19.909	0.000	–0.0086	1.472	0.236	Out:	BOD	–0.0284
16	0.9091	0.8264	0.7905	23.016	0.000	–0.0062	1.046	0.315	Out:	pH	0.3243
17	0.8998	0.8096	0.7779	25.513	0.000	–0.0168	2.815	0.104	Out:	Temp.	0.0071

equations. Selection of variable may be done using statistical software package like SPSS using backward elimination, forward selection or stepwise procedure. Forward selection starts with no independent variables in the equation and sequentially enters them. Backward elimination starts with all the variables in the equation and sequentially removes them. Stepwise selection of independent variables is a combination of backward and forward procedures. In SPSS package two removal criteria are available. The first is the minimum F value that a variable must have in order to remain in the equation. Variables with F values less than this F-to-remove (FOUT) are eligible for removal. Second criterion available is the maximum probability of F-to-remove (POUT) that a variable can have. Summary of Forward selection of variables at different steps with F, R^2 and Adj. R^2 values is presented in the Table 6.6. From the Table 6.6, it can be seen that through forward selection procedure at steps no. 1-5, Av.P, DO, INITWT, LnPL and Free CO_2 are selected. F-test reveals highly significant result in all cases. It is well known that R^2 (co-efficient of determination) is a measure of the goodness of fit of a particular model and R^2 of 0 does not necessarily mean that there is no association between variables. Instead, it indicates that there is no linear relationship. Changes in R^2 value can be observed at step 1-5. Highest R^2 value was observed in case of Av.P which indicates its importance in the growth of *L. rohita* in sewage fed ecosystem. All selected five variables together explain 80.96 per cent of the total variability in SGR of the test species. The sample R^2 tends to be an optimistic estimate how well the model fits the population. The model usually does not fit the population as well as it fits the sample from which it is derived. The adjusted R^2 attempts to correct R^2 to more closely reflect the goodness of fit of the model in the population. From above table adjusted R^2 is not found far away form R^2.

Table 6.6: Summary of Forward Selection of Variables at Different Steps with F, R_2 and Adj R_2 values

Step No.	*Variables*	*F*	*Sig. F*	*R^2*	*Adj R^2*
1.	Av. P	25.59	0.0000	0.4295	0.4127
2.	DO	34.22	0.0000	0.6747	0.6549
3.	INITWT	27.05	0.0000	0.7172	0.6907
4.	LnPL	23.78	0.0000	0.7542	0.7225
5.	Free CO_2	25.51	0.0000	0.8096	0.7779

Fitted equation $\hat{Y}(SGR) =$ 0.8084 + 0.4365 (Av.P) +.1339 (DO) +.1660 (Free CO_2)

0.0099 (INITWT) +.2209 (LnPL)
(R^2 =.8096; P<.0001)

Statistics of Variables not in the Equation

In the Table 6.4 in the column labeled Betaln is the standardised regression co-efficient that would result if the variable were entered into the equation at next step. The t test and level of significance are the hypothesis that the co-efficient is zero. The partial correlation coefficient with the dependent variable adjusts for the variables already in equation. From statistics calculated for *variables not in the equation*, one can decide what variable should be entered next.

Checking for Violation of Assumptions

In this particular case residuals are plotted against predicted variables and also against each independent variable. The distribution of residuals should be checked for normality.

Looking for Influential Points

Among the important influence measures is the leverage of a case. The leverage for a case describes the impact of the observed value of the dependent variable on the prediction of the fitted value.

Measures of Collinearity

Collinearity refers to the situation in which there is a high correlation between independent variables. The tolerance of a variable is commonly used as a measure of collinearity. Another important factor is the variance inflation factor (VIF), which is defined as the reciprocal of the tolerance. Tolerance and VIF values of the present analysis are furnished in Table 6.7. If the variance inflation factor increases so does the variance of regression co-efficient.

Eigen Values and Condition Indexes

Two useful tools for examining the collinearity of a data matrix are the eigen values of the scaled, uncentred cross product matrix and decomposition of regression variance corresponding to the eigen values. One can compare eigen values of the scaled, uncentered cross-product matrix to see if some are much larger than others. If this is the case, the data matrix is said to be ill conditioned. If a matrix is ill conditoined, small changes in the values of the independent or dependent variables may lead to large changes in the solution.

There are as many near-dependencies among the variables as there are large conditions indices. Table 6.8 shows the eigen values and condition indexes for the present example. It can be seen that the last two eigen values are much smaller than the rest. Their condition indexes are 9.889 and 12.517 respectively.

Variance Proportions

The variance of each of the regression co-efficient, including the constant, can be decomposed into a sum of components associated with each of eigen values. If a high proportion of variance of two or more coefficients is associated with the same eigen value, there is evidence for a near dependency. Consider the column Av.P. It is seen that 1.19 per cent of the variance of the co-efficient is attributable to the first eigen value, 4.98 per cent to the second and 0.53 per cent to the sixth. Looking at the last eigen values, it is seen that it accounts for 86 per cent of the variance of the constant, about 77 per cent of variance of DO and 29.6 per cent of initial wt. This indicates that these two variables are highly dependent. Since other independent variables have

Table 6.7: Measures of Collinearity-Tolerance and VIF

	Variables not in the Equation						
Variable	*Beta In*	*Partial*	*Tolerance*	*VIF*	*Min Toler*	*T*	*Sig T*
BOD	–0.053345	–.102754	0.706423	1.416	0.457801	556	0.5823
PH	098389	0.198336	0.773702	1.292	0.440655	1.090	2848
NH4_N	0.153854	0.235595	0.446454	2.240	0.405558	1.305	0.2020
ALKALINI	–0.062444	–0.110003	0.590882	1.692	0.462348	–.596	0.5558
CHLORIDE	0.039211	0.075993	0.715130	1.398	0.465374	0.410	0.6845
TEMP	–0.136778	–0.29747	1.900580	0.110	0.461298	–1.678	0.1041

Table 6.8: Measures of Collinearity-Eigen Values and Condition Indexes

Number	*Eigen value*	*Cond*	*Variance Proportions*					
		Index	*Constant*	*DO*	*AV.P*	*INITWT*	*LNPL*	*$FREECO_2$*
1	4.72407	1.000	0.00187	0.00262	0.01187	0.00496	0.00304	0.00713
2	0.69043	2.616	0.00005	0.00214	0.49784	0.03639	0.00065	0.05495
3	0.34185	3.717	0.00778	0.05508	0.44683	0.04482	0.02012	0.11737
4	0.16520	5.348	0.00070	0.03238	0.00054	0.35177	0.04376	0.54047
5	0.04831	9.889	0.12809	0.13849	0.03764	0.26562	0.92662	0.27384
6	0.03015	12.517	0.86152	0.76929	0.00526	0.29645	0.00580	0.00624

small variance proportions for the sixth eigen value, it does not appear that the observed dependencies are affecting their co-efficient.

Durbin–Watson Test (d)

Residual statistics and Durbin-Watson Test value is presented in Table 6.9. From the table, it is observed that Durbin-Watson test value was found to be 1.68. The possible value of the statistic range from 0 to 4. If the residuals are not correlated with each other, the value of d is close to 2. Values less than 2 mean that adjacent residuals are positively correlated. In the present case since d value is found 1.68 which is less than 2, it may be interpreted that adjacent residuals are positively correlated.

Table 6.9: Residuals Statistics

	Min	*Max*	*Mean*	*Std Dev*	*N*
*PRED	–0.1636	5.8122	1.4116	1.2261	36
*RESID	–1.1866	1.4071	0.0000	0.5946	36
*ZPRED	–1.2847	3.5890	0.0000	1.0000	36
*ZRESID	–1.8476	2.1908	0.0000	0.9258	36

Total Cases = 36; Durbin-Watson Test = 1.68172.

Interpretation and Validation of Model

The fitted equation with Av.P, DO, Initwt. LnPL and Free CO_2 as predictor variable are capable of explaining 80.96 per cent of total variability in the growth of *L. rohita*. Therefore, the model is a very good fit. All the identified variables play important role in the sewage fed ecosystem where no supplementary feed were provided to fish which survived on the natural growth of plankton that was again dependent on nutrient status of the pond.

References

Bhar Lalmohan, S. S. Walia and Roy, A. K., 2002. Forecasting Fish Production from Ponds (mimeo). Division of Forecasting Techniques, IASRI, Library Avenue, New Delhi-110012. p.30.

Boyd, C. E. 1979. Water quality in warm water fishponds. Aubum University Agricultural Experiment Station, Auburn, Alabama.

Brett, J. R. and Groves, T. D. D., 1979. Physicological energetics. In: W.S. Hoar, D.J. Randall and J.R. Brett (eds.) Fish physiology. Academic Press, New York, Vol. 8: p.279-352.

Chan, H. L., Garcia, M. C. and Leung, P., 2002. Long-term world projections of fish production and consumption. Report to the Food and Agriculture Organization of the United Nations.

Cuenco, M. L., Stickney, R. R. and Grant, W. E., 1985b. Fish bioenergetics and growth in aquaculture ponds: II. Effects of interactions among size, temperature, dissolved oxygen, unionized ammonia and food on growth of individual fish. *Ecological Modelling* 27:191-206.

Cuenco, M. L., Stickney, R. R. and Grant, W. E., 1985c. Fish bioenergetics and growth in aquaculture ponds: III. Effects of intraspecific competition, stocking rate, stocking size and feeding rate on fish productivity. *Ecological Modelling* 28:73-95.34

Dan Dam, A. A., 1987. Multiple regression analysis of accumulated data from aquaculture experiments: a rice-fish culture example. *Aquacult. Fish. Manage.* 21.

Delgado, C. L., Roseergrant, M., Meijer, S., Wada, N. and Ahmed, M., 2002a. Fish as food: Projections to 2020 within a global food model. Paper presented at the biennial meetings of the International Institute of Fisheries Economics and Trade, held in Corvalis, Oregon, July 10-14, 2002.

Delgado, Christopher L., Wada, N., Rosegrant, M. W., Meijer, S. and Ahmed, M., 2003. Supply and Demand in changing global Markets. IFPRI, Washington, D. C., and World Fish Centre, Penang.

Drapher, N. R. and Smith, H., 1981. Applied regression analysis, 2nd Edition, New York. pp. 407.

FAO, 1989. Aquatic production (1984-1986). FAO Fish. Circular, 815. Food and Agriculture Organization, Rome, Italy.

Gold, H. J. 1977. Mathematical modelling of biological systems–an introductory guidebook. John Wiley and Sons, New York.

Hair Jr., J.F., Anderson, R. E. and Tatham, R. L., 1987. Multivariate data analysis with readings. Second edition. Macmillan Publishing Company, New York.

Holmes, W. N. and Donaldson, E. M., 1969. Excretion, ionic regulation and metabolism. In: W.S. Hoar and D.J. Randall (eds.) Fish physiology. Academic Press, New York, Vol. 1: p.1-89.

Hopkins, K. D. and Cruz., E. M., 1982. The ICLARM-CLSU integrated animal fish farming project: final report. ICLARM Technical Reports 5. International Center for Living Aquatic Resources Management, Manila, Phillippines.

Ilstein, A., Hepher, B. and Teltsch, B., 1985a. Principal component analysis of interactions between fish species and the ecological conditions in fishponds: I. Phytoplankton. *Aquacult. Fish. Manage.* 16:305-317.

Karplus, W. J., 1983. The spectrum of mathematical models. *Perspectives in computing* 3(2):4-13.

Klemetson, S. L. and Rogers, G. L., 1985. Aquaculture pond temperature modelling. *Aquacultural Engineering* 4:191-208.

Krant, J., Motzkin, F. and Gordin, H., 1982. Modelling temperatures and salinities of mixed seawater fishponds. *Aquaculture* 27: 377-388.

Krant, J., Segal, I. and Dayan, A., 1985. A simple model for comparing energy exchange in warm soils and fishponds. *Energy in Agriculture* 4:25-39.

Meyer, D. I. and Brune, D. E., 1982. Computer modelling of the diurnal oxygen levels in a stillwater aquaculture pond. *Aquacultural Engineering* 1(4): 245-262.

Milstein, A., Hepher, B. and Teltsch, B., 1985a. Principal component analysis of interactions between fish species and the ecological conditions in fishponds: I. Phytoplankton. *Aquacult. Fish Manage,* 16:305-317.

Milstein, A., Hepher, B. and Teltsch, B., 1985b. Principal component analysis of interactions between fish species and the ecological conditions in fishponds: II. Zooplankton, *Aquacult. Fish. Manage.* 16:319-330.

Milstein, A., Hulata, G. and Wohlfarth, G. W., 1988. Canonical correlation analysis of relationships between management inputs and fish growth and yields in polyculture, *Aquacult. Fish. Manage.* 19:13-24.

Morita, S. K. 1977. An economic model for prawn pond production. *Proc. World Maricult. Soc.* 8:741-764.

Pauly, D. and Hopkins, K. D., 1983. A method for the analysis of pond growth experiments. *ICLARM Newsletter* 6(1):10-12.

Per Sparre, 1984. A programme for bioeconomic modelling on freshwater fish culture in India. F1: DP/IND/75/031, Field document, March, 1984 FAO, Rome, Italy.

Piedrahita, R. .H., 1988. Introduction to computer modelling of aquaculture pond ecosystems. *Aquacult. Fish. Manage.* 19:1-12.

Prain, M. and Milstein, A., 1988. Techniques for handling large pond and farm datasets. *Aquabyte* 1(2):4-5.

Riggs, D. S., 1963. The mathematical approach to physiological problems a critical printer. The MTT Press, Cambridge, Massachusetts.

Roy, A. K., Rout, M., Saha, P. K. and Datta, A. K., 1998. Influence of Environmental Factors on the SGR of *Catla catla* Grown in Sewage Ecosystem-Investigated with Multiple Regression Analysis. In: A. K. Aditya and P. Halder (Eds.), *Proc. of the National Seminar on Environmental Biology*. Daya Publishing House, New Delhi-35, pp. 87-93.

Roy, A. K., Saha; P. K., Datta, A. K., Satapathy, P. K., 2000. Regression model for *Catla catla* reared in sewage fed ponds In: B. B. Jana, R. D. Banerjee, B. Guterstam and J. Heeb (eds.) Waste Recycling and Resource Management in the Developing World, pp. 65-69. University of Kalyani, India and International Ecological Engineering Society, Switzerland held during 23-27, Nov., 1998 at Kolkata.

Roy, A. K., Satapathy, P. K. and Antony, A., 2006. Modelling in Aquaculture. In: A. K. Roy, A. Antony and N. Sarangi (eds.) *Bioinformatics and Statistics in Fisheries Research,* Vol-2. Central Institute of Freshwater Aquaculture, Kausalyaganga, Bhubaneswar-751002, India. Pp. 329-342.

Shell, E. W. 1968. Feeds and feeding of warm water fishes in North America. *FAO Fish. Rep.* 44(3):310-325.

Snedecor, G. W. and Cochran, W. G., 1980. Statistical Methods, Seventh edition. The Iowa State University Press, Ames, Iowa.

Spain, J. D., 1982. BASIC microcomputer models in biology. Addison-Wesley Publishing Co., London.

Spataru, P., 1976. Natural food of *Tilapia aurea* in polyculture with supplementary feed and intense manuring. Bamidgeh 28:57-63.

Svirezhev, Y. M., Krysanova, V. P. and Voinov, A. A., 1984. Mathematical modelling of a fishpond ecosystem. *Ecological Modelling* 21:315-337.

Swingle, H. S., 1968. Estimation of standing crops and rates of feeding fish in ponds. *FAO Fish. Rep.* 44(3):416-423.

Swingle, H.S. and Smith, E V., 1939. Fertilizers for increasing the natural food for fish in ponds. *Trans. Am. Fish. Soc.* 68:126-135.

Ursin, E., 1967. A mathematical model of some aspects of fish growth, respiration and mortality. *J. Fish. Res. Board Can.* 24:2355-2453.

Yamaguchi, R., 1981. A biophysical model of aquatic organism growth. Texas A &M University, College Station, Texas. 88 p. *Ph.D. Thesis.*

Modelling, Forecasting, Artificial Neural Network and Expert System in Fisheries
Editors: Ajit Kumar Roy & Niranjan Sarangi
Published by: DAYA PUBLISHING HOUSE (2009)
Pages 87–101

Chapter 7

Modelling of Multiobjective Freshwater Aquaculture

Pratap K.J. Mohapatra, Santosh K. Prusty and C.K. Mukherjee

Introduction

Aquaculture, the farming and husbandry of aquatic organisms, is important in the food economy of many nations (Pillay, 1979). Although much practical experiences have been studied in exotic species, there is much to be done for Indian major carps. The dynamics of the aquacultural system is very difficult to analyze because of the complex interactions among the factors affecting the system. Hence, the mathematical modelling of aquacultural system is still inadequate (Shell, 1983). Statistical knowledge can be applied to aquacultural experiments to unfold the relationship and understand the behavior of the system.

In India, there is a common practice to go for polyculture in ponds. Different species in varying stocking density affect the growth of the fish (Brett, 1979). The behavior of growth and other parameters were studied with different stocking densities and species combination (Jena, 2002). Fertilizer in fishpond plays an important role in semi- intensive polyculture, where a part of the food comes from the nature. Stocking density affects the natural food available per fish and level of supplementary feeding required (Moor, 1986; Hepher, 1988). In semi-intensive culture systems, the animal growth is dependent upon the consumption of endogenously available live food organisms and externally supplied feeds as a supplementary source of dietary nutrients (Tacon and De Silva, 1997). In such complexities, it is very difficult for the management to choose the level of input factors specifically when more than one input and outputs are considered. Techniques of design of experiments can be applied to aquacultural experiments to explore the significance of the factors and to decide the optimum levels of input factors to get maximum growth of fishes. Here the desirability function approach of optimization has been incorporated to optimize the outputs.

The statistical design of experiments refers to the process of planning the experiments so that appropriate data can be analyzed by statistical methods resulting in valid objective conclusion (Montgomery, 2001). Here the concept of design of experiment is used to find out the optimum level of the input factors in aquacultural process design as well as optimization of the process. In addition, the knowledge of statistical design helps in minimizing the number of experiments to be done, which is important in biological processes like aquaculture. The experiments for the present study were conducted in Central Institute of Freshwater Aquaculture (CIFA), Bhubaneswar, India. The analysis of variance for stocking density, fertilization, aeration, supplementary feeding upon fish growth has been carried out. Three experiments have been analyzed in the present study. For analysis of the results of different designs, the *Design-Expert 6.0* statistical tool has been used.

Experiment and Data Collection

A number of experiments were conducted at Central Institute of Freshwater Aquaculture (CIFA), Bhubaneswar, India, under various combinations of stocking densities, fertilizer and species. All the experiments were carried out by the scientists of CIFA in the ponds present in the institute. The general factorial designs are summarized in the Table 7.1. Broadly, we can classify the experiments of our analysis as follows:

Experiment I

The growth of carp species and biomass production in varied stocking density were tested. Six ponds of 0.04 hectare (20×20 m^2) were selected for the study. The culture period was up to 11 months. All the pre-stocking and post-stocking managements were done properly. Predatory and weed fishes were removed by applying bleaching powder at the rate of 525 kg/hectare-m. All the ponds were fertilized by cow dung at the rate of 15 ton/hectare/year, with equal split in bi-monthly schedule. Urea and SSP (Single Super Phosphate) were used as the inorganic fertilizers at the rate of 200 kg/hectare/year and 300 kg/hectare/year respectively for the whole period of culture. The stocking density was taken at three levels, *i.e.*, 2,000, 4,000 and 6,000 fingerlings/hectare. For each level of the factor, two replicates were taken. Single factorial design was conducted for this experiment and the best level of stocking density for optimum growth of fish was found out.

Experiment II

The effect of stocking density, fertilizer, supplementary feeding and aeration on the growth of carp species were experimented. Two designs have been incorporated here. In the first design, two factors–fertilization and stocking density–were each taken at three levels. In the second design, the aeration and stocking density have been considered. In the latter one, the stocking density was taken at three levels and aeration at two levels- with and without. General Factorial design was applied. In the second design, fertilization and supplementary feeding of the pond was carried out in each treatment. Cow dung, urea and SSP were used as the fertilizers. The rice bran and groundnut oil cake (1:1) were applied in the ponds as the supplementary feedings. Aeration had been done for 4-6 hours/day during nighttime by using aspirator aerator.

Table 7.1: Design Summary of the Factorial Experiments

Experiment No	*Type*	*Design*	*Factors*	*Levels*	*Replicates*	*Responses*
1	General Factorial Design		A: Stocking Density (numbers/ hectare)	2000 4000 6000	2	1. Survival (%) 2. Average harvested size (gram)
2	General Factorial Design	Design 1	A: Stocking Density (numbers/ hectare)	10,000 20,000 30,000	1	1. Survival (%) 2. Average harvested Size (gram) 3. Biomass Harvested (kg)
			B: Fertilization	Without (-1**) With (0**) With SF*(1**)		
		Design 2	A: Stocking Density (numbers/ hectare)	10,000 20,000 30,000	1	–do–
			B:Aeration	without with		
3	2^2 Factorial Design		A: No of Species	3 6	2	–do–
			B:Stocking Density (numbers/ hectare)	5,000 10,000		

*: SF: Supplemental Feeding; **: Coded factors.

Experiment III

In India, there is a practice to go for culturing multiple species in a pond to utilize the water depth efficiently. Generally, three or six species are taken. Three species- Rohu (*Labeo rohita*), Catla (*Catla catla*) and Mrigal (*Cirrhinus mrigala*)-are cultured in one pond. In a six-species combination three other species are added. They are Silver carp (*Hypophthalmichthys molitrix*), Common carp (*Cyprinus carpio*) and Grass carp (*Ctenopharyngodon idella*).

The fishes of these species are stocked at different stocking densities and species combination. Both the factors are taken at two levels. A 2^2 factorial design with two replicates has been done. $2 \times 2^2 = 8$ ponds have been selected for the experiment. The best combination of "stocking density" and "species combination" for optimum survival and growth of fish has been found out. In a three-species combination Rohu, Catla and Mrigal were taken in the ratio of 4: 3: 3. In a six-species combination, Catla, Rohu, Mrigal, Silver Carp, Grass Carp and Common Carp were taken in the ratio of 2 : 2.5: 2.0 : 1.5 : 0.5 : 1.5.

The necessary pre-stocking and post-stocking management practices were adopted in each treatment of every design. The data for growth and other water quality parameters have been recorded in each experiment. Rice bran and groundnut oil cake (1:1) were used as supplementary feeding. Cow dung, urea and single super phosphate were used as fertilizer. Aspirator aerator was used for aeration purpose.

Results and Discussion

Experiment I: General Factorial Design

ANOVA for Survival

The Analysis of variance (ANOVA) for survival for general factorial design with component of variations is shown in Table 7.2.

Table 7.2: ANOVA for Survival in Experiment 1

Source	Sum of Squares	DF	Mean Square	F Value	Prob > F	Comment
Model	34.44333	2	17.22167	1.359426	0.3799	not
A	34.44333	2	17.22167	1.359426	0.3799	significant
Pure Error	38.005	3	12.66833			
Cor Total	72.44833	5				

The "Model F-value" of 1.359426 implies the model is not significant relative to the noise. There is a 37.99 per cent chance that a "Model F-value" this large could occur due to noise.

The Box-Cox Plot (Figure 7.1) shows that the value of λ is equal to one. This signifies that the model does not need transformation. The points in the normal plot of residuals (Figure 7.2) follow a straight-line, signifying that the sample is normally distributed. The I -beams of the one-factor plot (Figure 7.3) shows that there is no significant effect of stocking density on the survival percentage of fish population. While analyzing the model for mean weight of fish at the time of harvest, the "Prob >F" value of 0.0108 shows that the stocking density has a significant effect on the average harvested size of fish (Figure 7.4). The model has been optimized keeping the stocking density within the limit of 2,000 to 6,000 numbers/hectare. To get maximum survival and maximum mean weight, the optimum stocking density was found to be 2,000 fishes per hectare. The desirability plot (Figure 7.5) shows that there is maximum desirability of 83.60 per cent in case of 2,000 numbers/hectare stocking density.

Experiment II

Two designs have been incorporated here. The effect of fertilizer, stocking density, supplemental feeding and aeration upon the survival and growth of fish has been analyzed by general factorial design method.

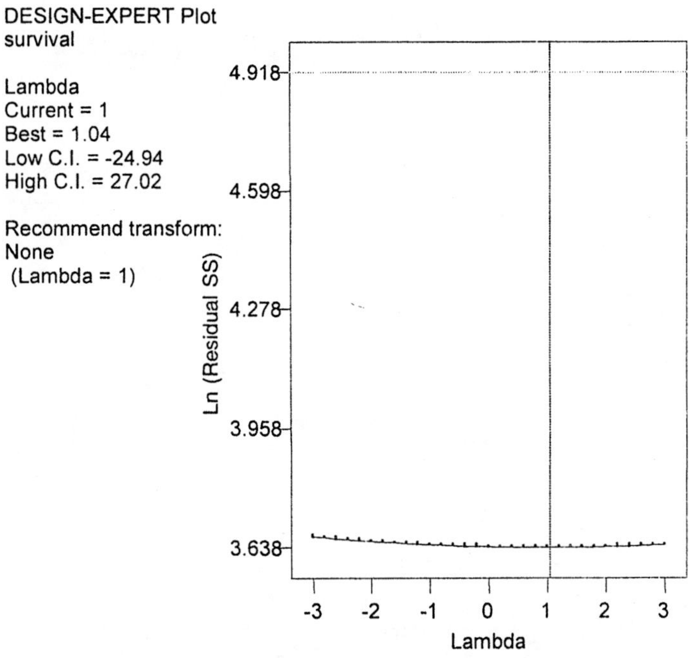

Figure 7.1: Box-Cox Plot for Power Transformation

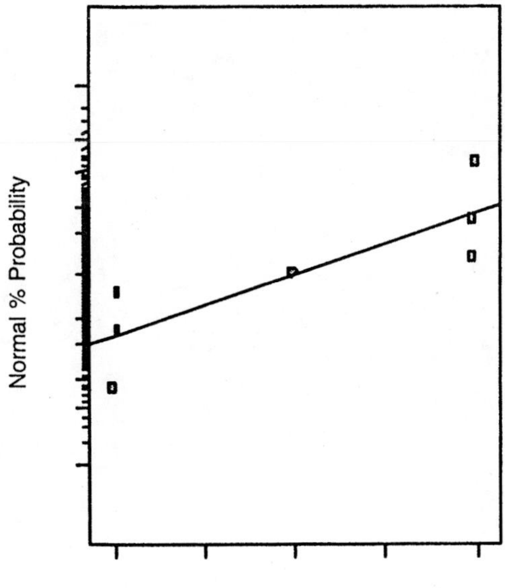

Figure 7.2: Normal Plot of Residual

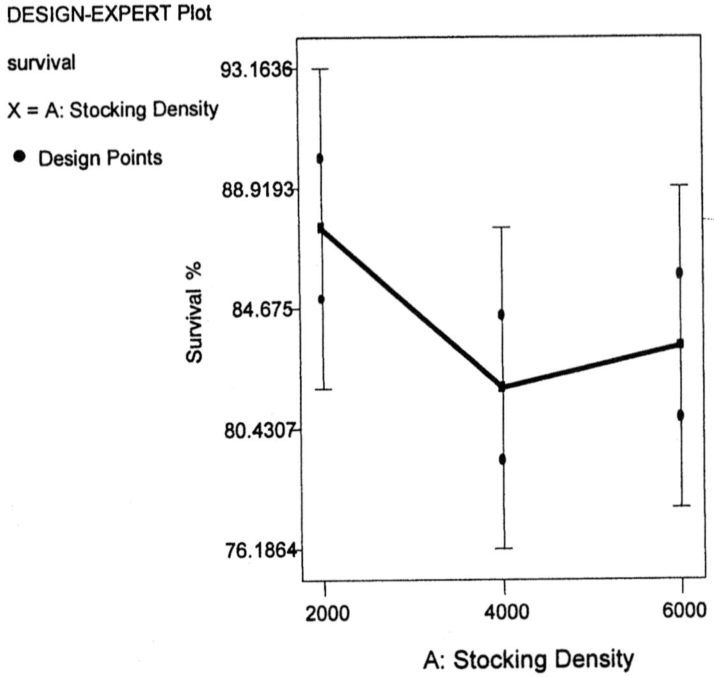

Figure 7.3: One Factor Plot for Survival

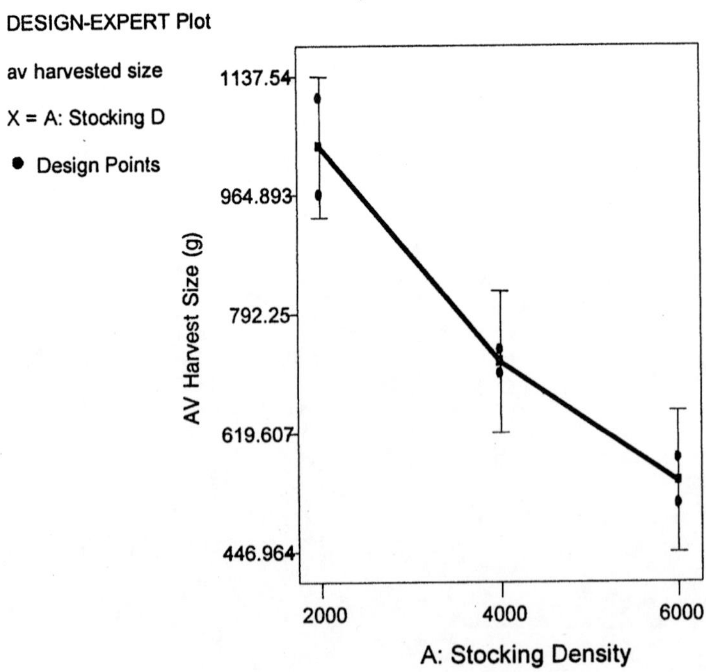

Figure 7.4: One Factor Plot for Mean Weight at Harvest

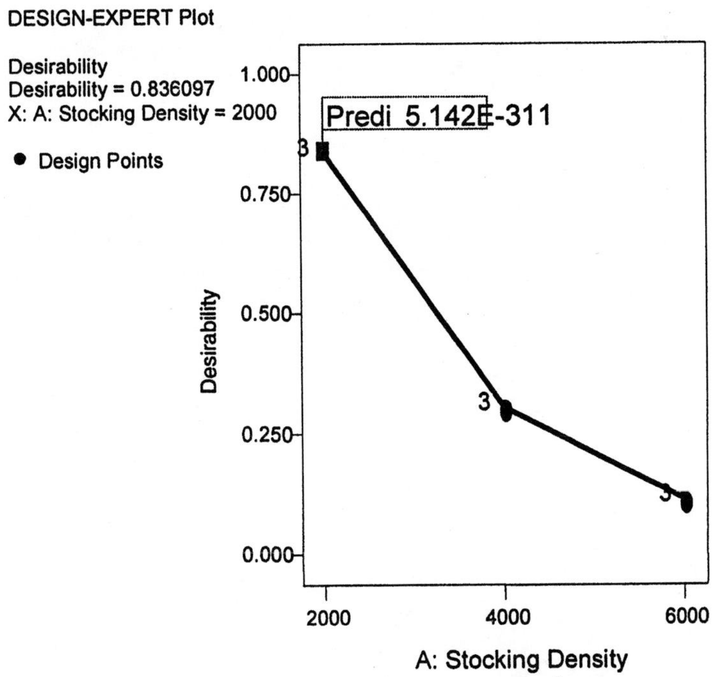

Figure 7.5: Desirability Plot of Optimization

Design 1

The ANOVA table (Table 7.3) shows that the model is significant. And the "Prob > F" value is less than 0.005 for each of the model terms- stocking density and fertilizer application. Therefore, both the model terms are significant. No such particular pattern of the "residual Vs run plot" shows that the independence of constant variance assumption holds good. The normal plot of residuals shows that residuals follows normal distribution.

Table 7.3: Analysis of Variance (ANOVA) for Survival in Experiment II

ANOVA Source	*Sum of Squares*	*DF*	*Mean Square*	*F Value*	*Prob > F*	
Model	2716.187	4	679.0467	74.83982	0.0005	significant
A	841.5267	2	420.7633	46.37362	0.0017	
B	1874.66	2	937.33	103.306	0.0004	
Residual	36.29333	4	9.073333			
Cor Total	2752.48	8				
Std. Dev.	3.012197		R-Squared	0.986814		
Mean	67.1		Adj R-Squared	0.973629		
C.V.	4.489117		Pred R-Squared	0.933247		
PRESS	183.735		Adeq Precision	24.97224		

As the all R^2 values are high, hence the experimental error is less. Hence, the model can be utilized for the design purpose. The "Pred R-Squared" of 0.933247 is in reasonable agreement with the "Adj R-Squared" of 0.973629 (Table 7.3). "Adeq Precision" measures the signal to noise ratio. A ratio greater than 4 is desirable. The ratio of 24.97224 indicates an adequate signal. This model can be used to navigate the design space. The PRESS value is less compared to SS_T (Table 7.3). Hence the small value of PRESS indicates that the model is likely to be a good predictor. The interaction plot (Figure 7.6) shows that the combination of fertilization and lower stocking density gives better survivality.But supplemental feeding with addition to fertilizer application does not show significant increase in survivality of the fish.

The model has been analyzed for average size of the fish at the time of harvest. The Box–Cox Plot (Figure 7.7) recommends for the inverse square root transformation of the model. The ANOVA for this design shows that value of "Prob > F" less than 0.0500, hence the model is significant.

All the model parameters show that the model is likely to be a good predictor and can be used for design purpose. The R^2 value of the model 0.991683 shows the goodness of the model. The interaction plot (Figure 7.8) signifies that the maximum size of the fish occurs at 10,000 numbers/hectare stocking density and fertilizer application with supplemental feeding.

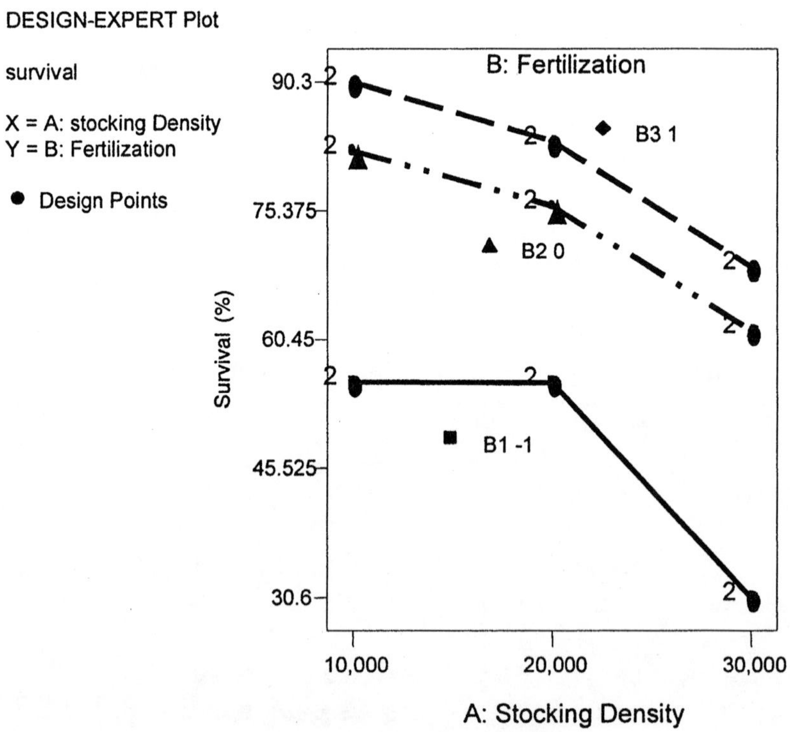

Figure 7.6: Interaction Plot between Fertilization and Stocking Density upon Survival

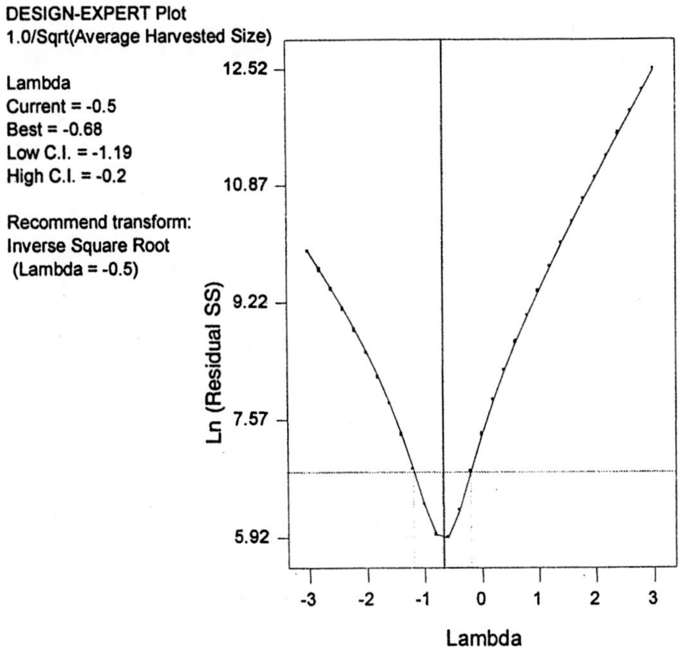

Figure 7.7: Box-Cox Plot for Factorial Design for Average Size of Fish at Harvest

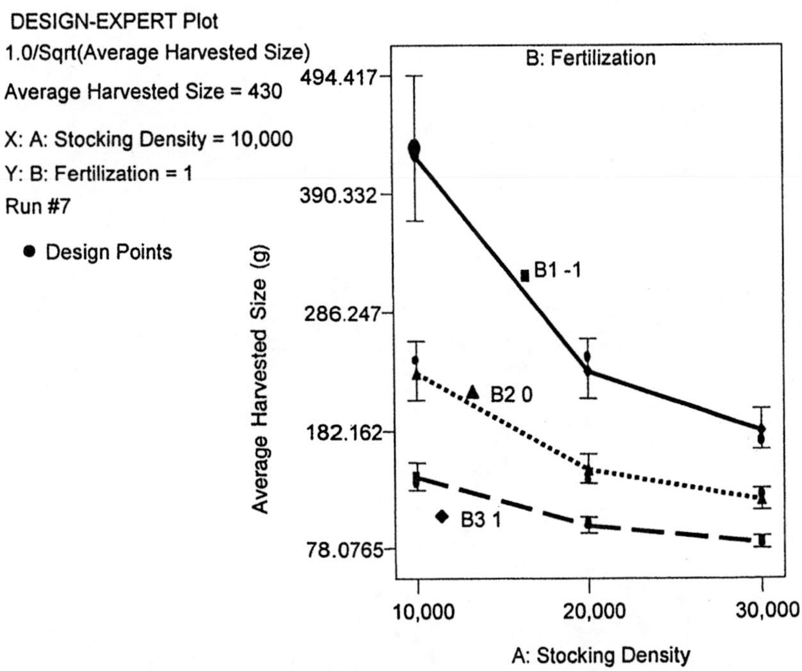

Figure 7.8: Interaction Plot between Fertilization and Stocking Density upon Average Size of Fish at Harvest

Again, the model analyses the third response- biomass harvested. The model does not need any transformation. The model is a good predictor with R^2 value of 0.990431.The ANOVA for biomass-harvested shows that there is significant effect of both the factor on the total biomass harvested. The interaction graph (Figure 7.9) shows that the maximum biomass can be harvested at 20,000 numbers/hectare stocking density and fertilizer application with addition to supplemental feeding.

The model was optimized with the objective of getting maximum responses. However the "total biomass harvested" was given primary weight compared to" survival" and "average harvested size". Seven numbers of solutions were found (Table 7.4). Combination of 10,000 numbers/hectare stocking density with fertilizer application and supplementary feeding gives biomass harvested at 155.2 kg, a maximum average harvested fish size of 430 gm and a survival of 90.3 per cent. In the second solution, though the maximum biomass harvested is 164 kg, the average harvested size is significantly smaller of a 246 gm.

Design 2

In this design, the analysis has been done to study the effect of aeration upon the responses- survival of fish, average harvested size of fish and biomass harvested. The ANOVA for each of the responses was analyzed and it was found that all the

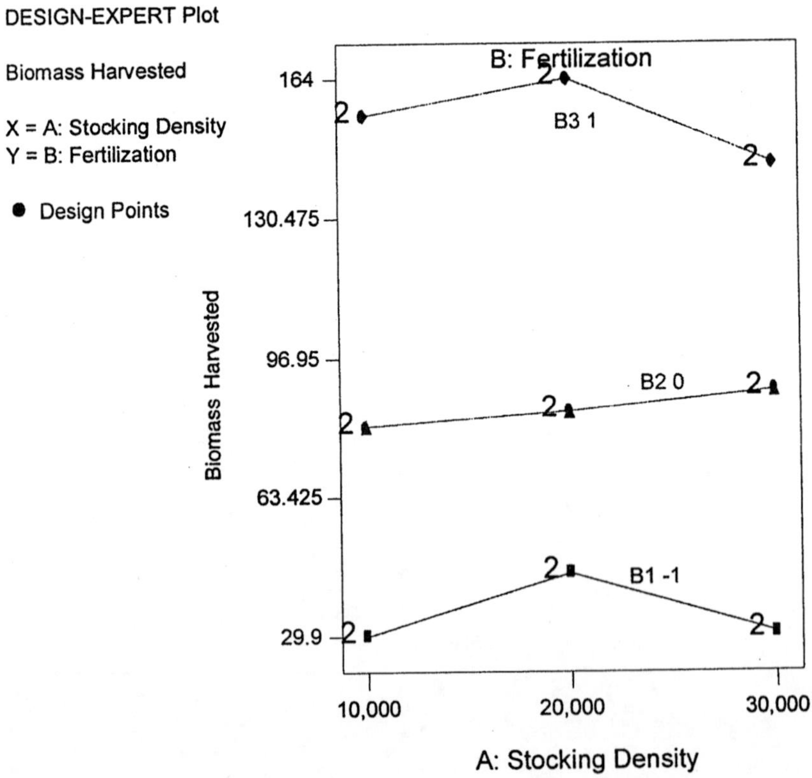

Figure 7.9: Interaction Plot between Fertilization and Stocking Density upon Total Biomass Harvested

three responses *e.g.* survival, average size of fish at harvest, total biomass harvested have been affected by stocking density and aeration. However, the interaction effect between the two factors significantly affects the total biomass harvested.

Table 7.4: Solution Table for Optimization of Process for Design 1

Solutions Number	*Stocking Density*	*Fertilization*	*Survival*	*Average Harvested*	*Biomass Harvested Size*	*Desirability*
1	10,000	1	90.3	430	155.2	0.969619
2	20,000	1	83.4	246	164	0.915625
3	30,000	1	68.9	174	143.5	0.72834
4	10,000	0	82.3	244	80.4	0.583351
5	20,000	0	76	138	84	0.504641
6	30,000	0	61.5	125	89	0.450774
7	20,000	-1	55.4	101	44.9	0.189762

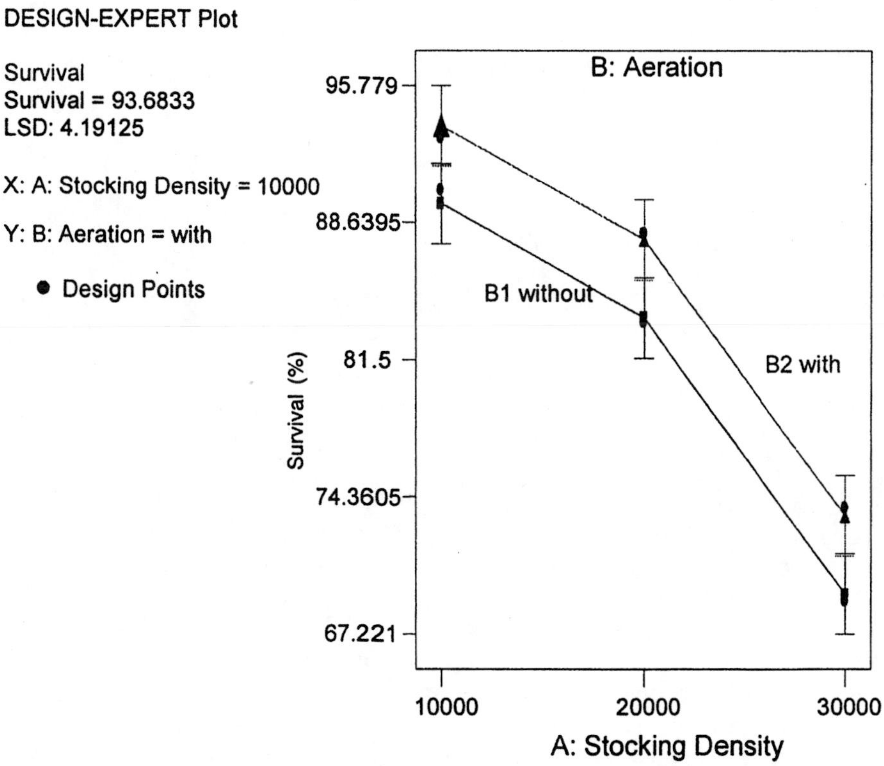

Figure 7.10: Interaction Plot between Aeration and Stocking Density upon Survival

The aeration has been found to be significant for the three responses. However, 10,000 numbers/hectare stocking density with aeration gives maximum average harvested size (Figure 7.11). However, in case of biomass production, stocking density of 20,000 numbers/hectare and aeration gives better result (Figure 7.12). While the model was optimized giving same weight to average harvested size and biomass production, it has been found that stocking density of 20,000 numbers/hectare with supplementary aeration give maximum biomass production and good average harvested size.

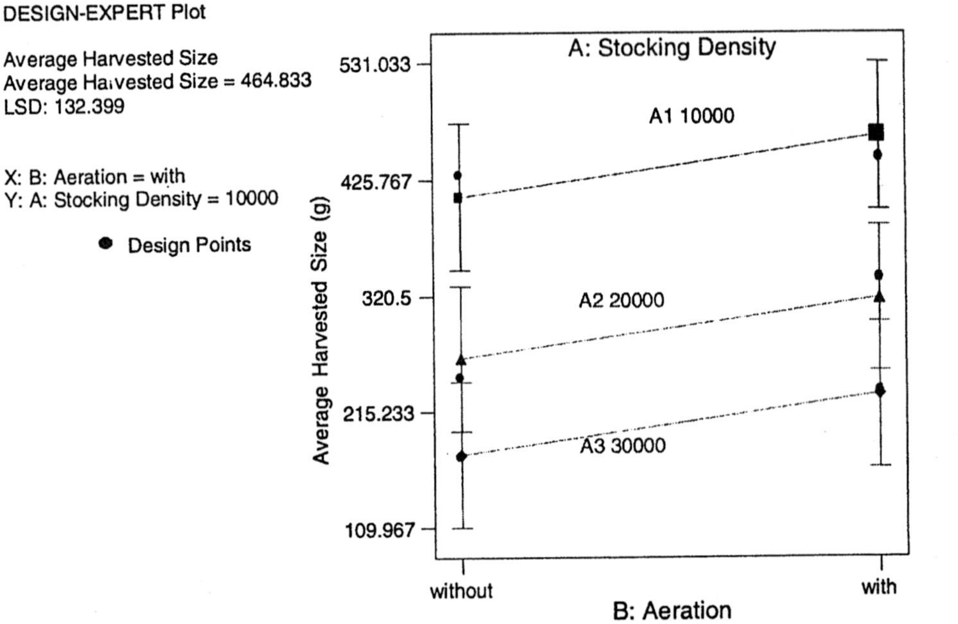

Figure 7.11: Interaction Plot between Aeration and Stocking Density Upon Average Harvest Size

Experiment III

Different stocking densities with different species combinations have an effect on the growth of the fish. Here a 2^2 factorial design of experiment analyses the effect on the responses- survival, average harvested fish size and biomass harvested. The ANOVA for survival of the fish shows that the model is not significant- the factors are not having any significant effect on the survival of fish. The I-beams of the interaction plot (Figure 7.13) shows that the factors are not having significant effect on fish survival. However, at six-species combination, stocking density has little effect on the survival. At three-species combination, stocking density has no significant effect on survival of fish. The ANOVA for the average harvested fish size shows that the model is significant. The interaction plot (Figure 7.14) shows that the 5,000 numbers/hectare stocking density with six species combination gives the maximum "average harvested fish size". The analysis of variance was done for total biomass harvested.

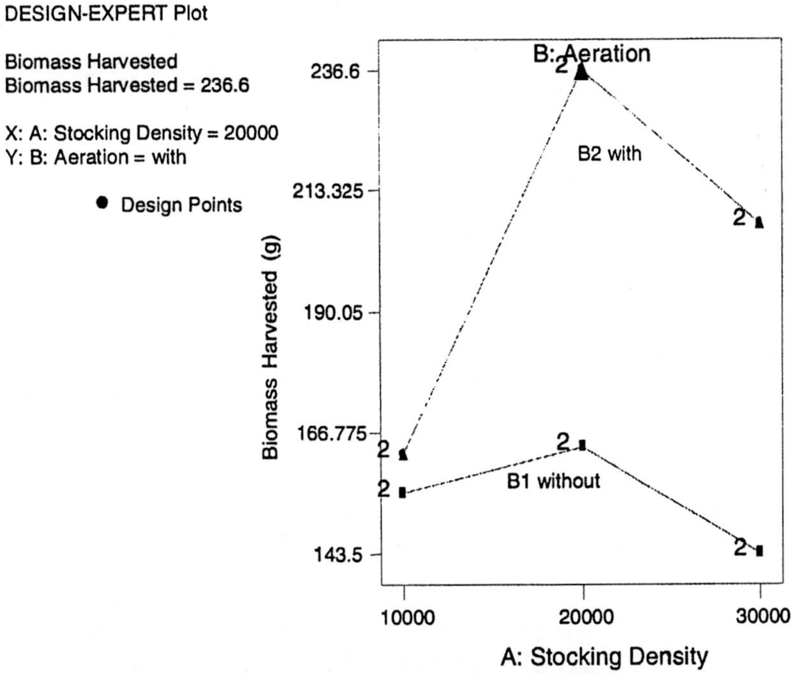

Figure 7.12: Interaction Plot between Aeration and Stocking Density upon Biomass Harvested

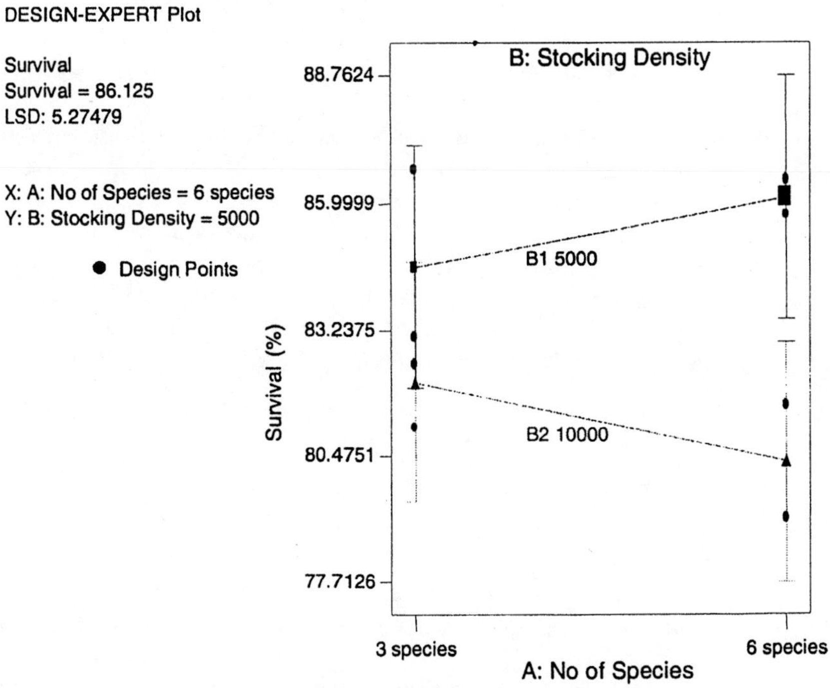

Figure 7.13: Interaction Plot for Survival of Fish

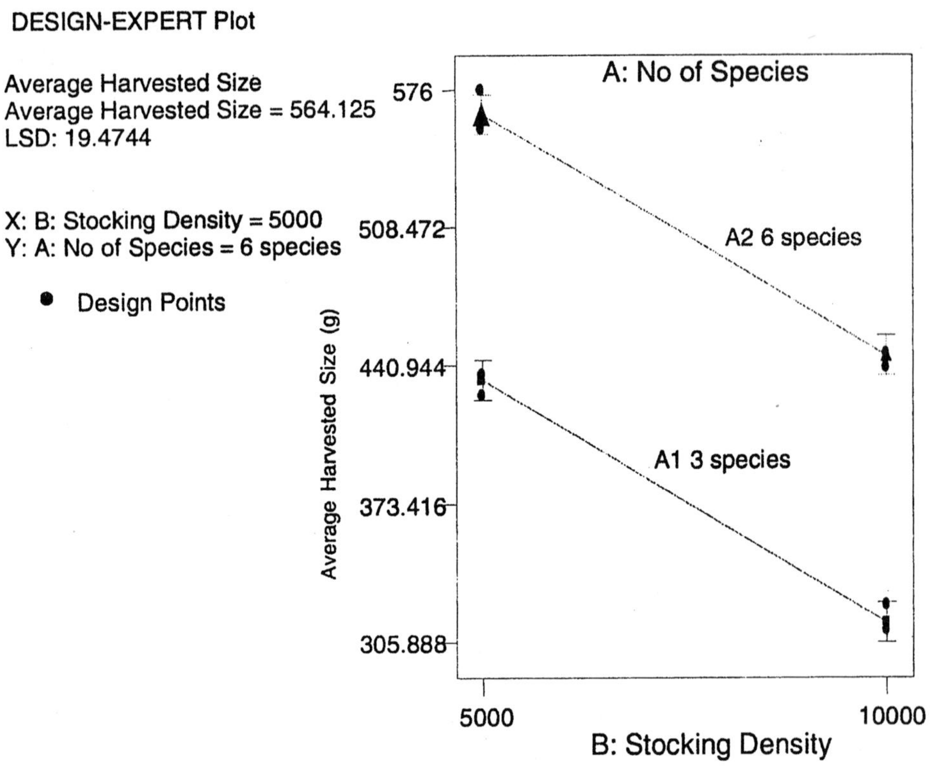

Figure 7.14: Interaction Plot for Average Harvested Fish Size

The model shows that the factors have significant effect on the response. It seems that, there exists interaction effect between the factors. It was found that the combination of 10,000 stocking density with six species culture gives the maximum biomass production. The model was optimized to get maximum responses. However, the biomass production was given primary weight than average harvested size. The solution (Table 7.5) shows that, a combination of 5,000 numbers/hectare stocking density with 6 species culture gives better desirability than 10,000 numbers/hectare stocking density. It depends on the management to choose the solution.

Table 7.5: Optimization Solution Table for Stocking Density and Species Combination

Solutions Number	*No of Species*	*Stocking Density (No/ha)*	*Survival (%)*	*Average Harvested Size (g)*	*Biomass Harvested (kg)*	*Desirability*
1	6 species	5000	86.12	564.125	195.325	0.628624
2	6 species	10000	80.35	445.875	285.225	0.49731
3	3 species	5000	84.63	433.875	146.45	0.174482
4	3 species	10000	82.1	315.625	208.85	0.135624

Conclusion

The maximum survival and mean weight were found to occur in the case of minimum stocking density of 2,000 numbers/hectare. The harvested size of the fish and survival decreases with an increase in stocking density. The second experiment shows that for a fully fertilized pond with supplemental feeding, the stocking density has negative effect on survival of fish and the harvested mean size. However, it has a positive effect on biomass-harvested upto certain limit. However, a lower value of 10,000 numbers/hectare stocking density with fertilizer and supplementary feeding gives optimum biomass harvest and maximum size of the fish. In the semi intensive ponds, aeration has a positive effect on all the response variables. There is existing an interaction effect between aeration and stocking density upon the biomass harvested. Hence in case of aerated pond, the stocking density has a significant effect on the biomass harvested. The average size of the fish was drastically reduced and the biomass production was significantly increased when the stocking density varied from 10,000 to 20,000 numbers/hectare. After 20,000 the stocking density has a negative impact on both the responses. While going for multiple species culture, the six species combination with 5,000 numbers/hectare gives higher desirability than, stocking with 10,000 numbers/hectare.

References

Boyd, C. E. and Egna, H. S., 1997. Dynamics of Pond Aquaculture,325-375,CRC Press, New York.

Boyd, C. E., 1990. Water Quality in Ponds for Aquaculture, Auburn University, Auburn, Albama.

Brett, J. R., Groves, T. D. D., 1979. Physiological Energetics. In: Hoar, W.S., Randall, D.J.,(Eds.), *Fish Physiology*, Academic Press, NY, 8:279-347.

Jena, J. K., Ayyappan, S., Aravindakshan, P. K., Dash, B., Singh, S. K. and Muduli, H. K., 2002. Evaluation of production performance in carp polyculture with different stocking densities and species combinations. *J. Appl. Ichthyol.* 18:165-171.

Montgomery, D. C., 2001, Design and Analysis of Experiments. 11-13, John Wiley and Sons, Inc., New York.

Moor, L. B., 1986. Input of organic materials into Aquacultural Systems. Emphasis on feeding semi-intensive systems. *Aquacultural Engineering*, 5:123-133.

Morita, S. K., 1977. An Economic Model of Prawn Pond Production. Proc.World Maricult. Soc., 8:741-764.

Pillay, T. V. R., 1979. The State of Aquaculture, pp 1-10,Fishing New Books Ltd., Farnham, England.

Shell, E. W., 1983. Fish Farming Research, Albama Agricultural Experiment Station. Auburn University; Auburn, Albama.

Tacon, A. G. J. and De Silva, S. S., 1997. Feed preparation and feed management strategies within semi-intensive fish farming systems in the tropics. Aquaculture. 151:379–404.

Modelling, Forecasting, Artificial Neural Network and Expert System in Fisheries
Editors: Ajit Kumar Roy & Niranjan Sarangi
Published by: DAYA PUBLISHING HOUSE (2009)
Pages **102–111**

Chapter 8

A Revisit to Building up Models in Fishery Research

Satyabrata Pal and Subhabaha Pal

Introduction

Human endeavor to translate the interplay of actions and interactions of the factors governing the natural phenomena in a more explicit form, easily comprehensible, so as to generate nice interpretation useful for decision making activities, is eternal. This endeavor has given birth to the concept of models, which serve as scientific representations of the behavior of such phenomena. Thus a model is a mechanism to develop the relationship, in mathematical terms, between the interplaying factors (variables) in a natural phenomenon/scientific experiment/system. More specifically, a model is an expression in mathematical form between the output (dependent) variables and the input (independent) variables revealing the truth behind the system. In the process to excavate the truth, the realization of uncertainty and variation governing such phenomenon/system led him to enrich the content/inferential power of the model by incorporating appropriate elements with respect to these in the model. Most of the models used in present days contain such elements embedded in the fabric of these.

This article deals with the fishery models involving growth and time variables. The growth variable considered is the length of fish. Along with the celebrated von Bertalanffy (VB) growth model, other parametric models, namely, Putter growth curve, Gompertz growth model, Richards growth model, and Logistic growth model have been explored in order to have a comparison of their powers to represent/predict the observed data with maximum closeness. A real–life data set has been employed in order to judge their predictive powers.

Growth Models

A dynamic model based on length determines the body size (length) of fish as a function of age (time). The mathematical model developed for individual growth by von Bertalanffy (VB) is well known to conform to the observed growth for most of the fish species and it is regarded as a cornerstone in fishery biology. The expression of the model is given below.

$$L_t = L_\infty \times [1-\exp(-K \times (t-t_{0)})], \quad (1)$$

where,

L_∞: Length of the fish at the end of life-time (in fact, it is the maximum attainable length),

L_t: Length of fish observed at time t,

K: Value of the growth parameter (curvature parameter) or growth rate, and

t_0: The time point when the fish has zero length.

Biologically, this (t_0) is not meaningful as the growth begins at hatching when the larva has already a certain length.

The above parameters are to be estimated from a given set of data.

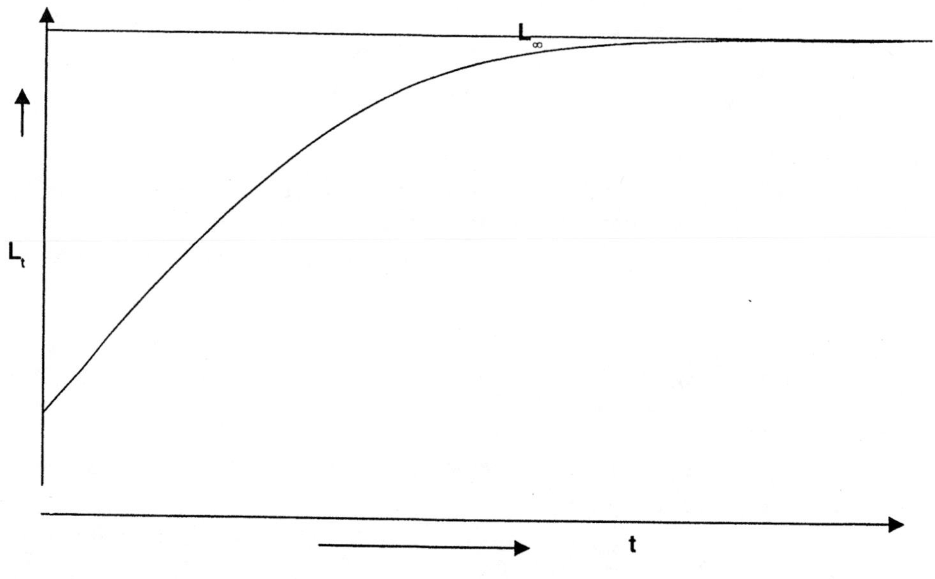

Figure 8.1 : von Bertalanffy Graph Plot

In the following, some simple explanations are presented.

If t = 0, then $L_0 = L_\infty \times [1-\exp(-K \times t_0)]$, meaning thereby that the length of the larva has a certain value at birth. The growth of fish begins at hatching when the larva has a certain length.

It is a fact that the growth rate of fish varies with age, as age increases growth rate decreases.

The growth feature is represented by the following expression:

$$\text{Let } \Delta L(t) = L(t+\Delta t) - L(t); \text{ then, } \Delta L(t)/\Delta t = a + b\,L(t) \qquad (2)$$

The VB growth equation can also be written as,

$$\Delta L(t)/\Delta t = K \times [L_\infty - L(t)] \qquad (3)$$

where by comparing (3) with (4) we get,

$$K = -b, \text{ and } L_\infty = -a/b \qquad (4)$$

An easy way to find out the values of K and L_∞ from the data is as follows:

Define

$$L^+(t) = \frac{[L(t+\Delta t) + L(t)]}{2}$$

Plot $\Delta L(t)/\Delta t$ versus $L^+(t)$.

A best fitting straight line can be drawn by hand, OR, a regression line can be set up between the dependent variable, $\Delta L(t)/\Delta t$ and the independent variable, $L^+(t)$.

When a straight line is drawn by hand, the gradient, b, of the line gives the value of –K. Further, from the intercept, a, of the straight line the value of L_∞ can be obtained from the relation $L_\infty = -a/b$. The physical interpretation of the parameter L_∞ is that at L_∞ the value of the quantity, $(\Delta L(t)/\Delta t)$ or rate of growth, is zero.

Regression Line Approach

The regression coefficient estimated from the data $[(\Delta L(t)/\Delta t, L^+(t)]$ gives the value of–K. L_∞ can be estimated from the intercept.

Von Bertalanffy presented the following procedure for estimation of the constants, L_∞, K, t_0.

Procedure of Estimation of the Constants

$$\text{Recall (1), } -\ln(1 - L(t)/L_\infty) = -K \times t_0 + K \times t \qquad (5)$$

The above equation suggests that $\ln(1 - L(t)/L_\infty)$ is a linear equation in t.

Firstly, an estimate of L_∞ is inserted as an input. Any one of the following considerations may act as guide:

(*a*) In small sample, length of the largest fish may be used as an estimate of L_∞.

(*b*) In large sample, average length of ten largest fishes may be used as an estimate.

(*c*) The value of L_∞ as obtained from the regression equation 3.

(*d*) Select that value of L_∞ around the value obtained in (c), which minimizes the R- square value.

From (5) it is clear that, $-\ln(1 \times L(t)/L_\infty)$ is a linear regression function in t. Thus from the estimated regression coefficient and the intercept, the values of K and t_0 can be obtained. In fact, the value of the regression coefficient is equal to K, and the value of the intercept = $-K \times t_o$ or t_o = –intercept/K*, being the value of K obtained from the given data set.

Apart from the above deterministic fitting of the VB model equation on a real-life data set, the parameters involved in the equation have also been estimated by employing non-linear regression method in this paper. The other important and useful growth models, *viz*, Putter, Gompertz, Richard's, and Logistic have also been applied on the same data set. Parameter estimation in all the cases has been done invoking non-linear optimization. A glimpse at the model equations of the above-mentioned five models is given below.

VB growth model: $L_t = L_¥ * [1- \exp(-K^* (t-t_{o)})],$ (1)

Putter growth model: $L_t^{(p)} = L_\infty^{(p)} * [1-\exp(-k^{(p)}*t)]^\wedge(1/3)$ (2)

Gompertz growth model: $L_t^{(g)} = L_\infty^{(g)} * \exp(-\exp(-k^{(g)}*t)),$ (3)

Richards growth model: $L_t^{(R)} = L_\infty^{(R)} * [1+b^{(R)} * \exp(-k^{(R)} *t))^\wedge(-1/b)$ (4)

Logistic growth model: $L_t^{(L)} = L_\infty^{(L)} * ((1+\exp(-k^{(L)}*t)^\wedge(-1))$ (5)

The data set comprised of the mean values of 10 fish samples ((*Puntius sophore* (Ham.)) on lengths L and 12 observations corresponded to each observation being recorded after a period constituting 4 week-interval. Ultimately, weekly averages (extending from the 4th week to the 48th week) are considered for development of the growth models.

Data Set

Table 8.1: The Following Table Gives the Record of 12 Values on Length of Fish

Serial No. (t)	1	2	3	4	5	6	7	8	9	10	11	12
Length [L (t)] (cm)	5.67	5.72	6.22	6.75	7.13	7.72	8.10	8.20	8.22	8.38	8.63	8.88

Recalling the VB equation:

$$L(t) = L_\infty \times [1-\exp(-K \times (t-t_{o)})]$$ (6)

The procedure is to obtain a trial value of $L_¥$ by considering the regression equation of $\Delta L(t)/\Delta t = a + b \times L^+(t)$. The detailed procedure for obtaining the trial value is given in Table 8.2.

Results and Conclusion

From the Table 8.2, the regression equation of $\Delta L(t)/\Delta t$ on $L^+(t)$ is:

$\Delta L(t)/\Delta t = .7426-0.0603\ L^+(t)$ and K= 0.0603 and $L_\infty = .7426/.0603 = 12.31$. A trial value of L_∞ is taken as 12. For many trial values of $L_¥$ near 12, it can be found that for $L_\infty = 11$, maximum value of $R^2 = 0.9743$ is obtained. The procedure of fitting $L_\infty = 11$ is described in Table 8.3.

The VB equation is then fitted applying non-linear optimization technique. Also the other four models were fitted by employing non-linear optimization. The predicted values are given in Table 8.4.

Table 8.2: Output of Parameter Estimates

t	L (t)	ΔL(t)	ΔL (t)/Δt = ΔL (t), Δt=1	L+(t)	[ΔL (t)]²	[L+(t)]²	ΔL (t)* L+(t)	SS [ΔL (t)]	SS [L+(t)]	SP [L+(t), ΔL(t)]	Corr. Coeff. [L+(t), ΔL(t)]	Reg. Coef. [ΔL(t), L+(t)]	Inter-cept	L (∞)
1	5.675													
2	5.725	0.05	0.05	5.7	0.0025	32.49	0.285							
3	6.225	0.5	0.5	5.973	0.25	35.70063	2.9875							
4	6.75	0.525	0.525	6.4875	0.275625	42.08766	3.4059							
5	7.125	0.375	0.375	6.9375	0.140625	48.12891	3.4059							
6	7.725	0.6	0.6	7.425	0.36	55.13063	4.455							
7	8.1	0.375	0.375	7.9125	0.140625	62.60766	2.9671							
8	8.2	0.1	0.1	8.15	0.01	66.4225	0.815							
9	8.225	0.025	0.025	8.2125	0.000625	67.44516	0.20531							
10	8.375	0.15	0.15	8.3	0.0225	68.89	1.245							
11	8.625	0.25	0.25	8.5	0.0625	72.25	2.125							
12	8.875	0.25	0.25	8.75	0.0625	76.5625	2.1875							
	89.625	3.2	3.2	82.35	1.3275	627.7156	23.28	0.3965	11.2135	-0.676	-0.320	-.0603	.7426	12.31
			0.2909	7.4863										

Table 8.3: Output of Various Estimates

t	t^2	L(t)	Z(t)	$[Z(t)]^2$	t*Z(t)	SS(t)	SS [Z(t)]	SP [Z(t),t]	Reg. Coeff. (K)	Intercept	t (zero)	Estd. L (t)	Error	Sq. Error	R–sq Error
1	1	5.675	0.725	0.526	0.725							5.577	0.097	0.009	
2	4	5.725	0.734	0.540	1.469							6.026	–0.031	0.09	
3	9	6.225	0.834	0.696	2.503							6.438	–0.213	0.045	
4	16	6.75	0.950	0.904	3.803							6.815	–0.065	0.004	
5	25	7.125	1.043	1.088	5.216							7.161	–0.036	0.001	
6	36	7.725	1.211	1.467	7.269							7.479	0.245	0.060	
7	49	8.1	1.333	1.777	9.332							7.770	0.329	0.1084	
8	64	8.2	1.368	1.872	10.94							8.037	0.162	0.026	
9	81	8.225	1.377	1.896	12.395							8.283	–0.058	0.003	
10	100	8.375	1.432	2.052	14.328							8.507	–0.132	0.017	
11	121	8.625	1.532	2.349	16.861							8.714	–0.089	0.007	
12	144	8.875	1.644	2.703	19.729							8.9032	–0.028	0.0007	
78	650		14.189	17.875	104.582	143	1.097	12.351	0.08637	0.621	–7.18			0.376	0.9743
6.5	54.16		1.182												

In the above table,

$Z(t) = \ln(1 - L(t)/L_x)$, the regression coefficient of Z(t) on t is 0.8637 or K = 0.8637, and to = –7.18 taking $L_x = 11$.

In fact, with the above mentioned parameter-values, the value of R^2 is maximised.

Table 8.4: Comparison of Estimates of Various Models

Week (t)	*Observed Fish Length (cm)*	*VB (Deterministic) Fish Length (Estimated)*	*VB Fish Length (Estimated) (Non-linear Fit)*	*Putter Fish Length (Estimated)*	*Gompertz Fish Length (Estimated)*	*Richards Fish Length (Estimated)*	*Logistic Fish Length (Estimated)*
1	5.675	5.577	5.416352	4.7521	4.38419	5.43457	5.333128
2	5.725	6.026	5.947485	5.847171	5.430293	5.92194	5.824057
3	6.225	6.438	6.418323	6.540653	6.29399	6.385871	6.294481
4	6.75	6.815	6.83571	7.03891	6.968545	6.816932	6.736228
5	7.125	7.161	7.205714	7.418306	7.475511	7.208321	7.143235
6	7.725	7.479	7.533715	7.717025	7.846537	7.556161	7.511718
7	8.1	7.770	7.824479	7.957367	8.113136	7.859359	7.840066
8	8.2	8.037	8.082235	8.153666	8.302288	8.119139	8.128536
9	8.225	8.283	8.31073	8.315756	8.435322	8.338419	8.378835
10	8.375	8.507	8.513286	8.450707	8.528326	8.521167	8.593676
11	8.625	8.714	8.692847	8.563782	8.593073	8.671844	8.776378
12	8.875	8.9032	8.852023	8.659005	8.638021	8.794975	8.930524
	R-square	0.9743	0.976983	0.9135661	0.85304	0.9809369	0.9754225

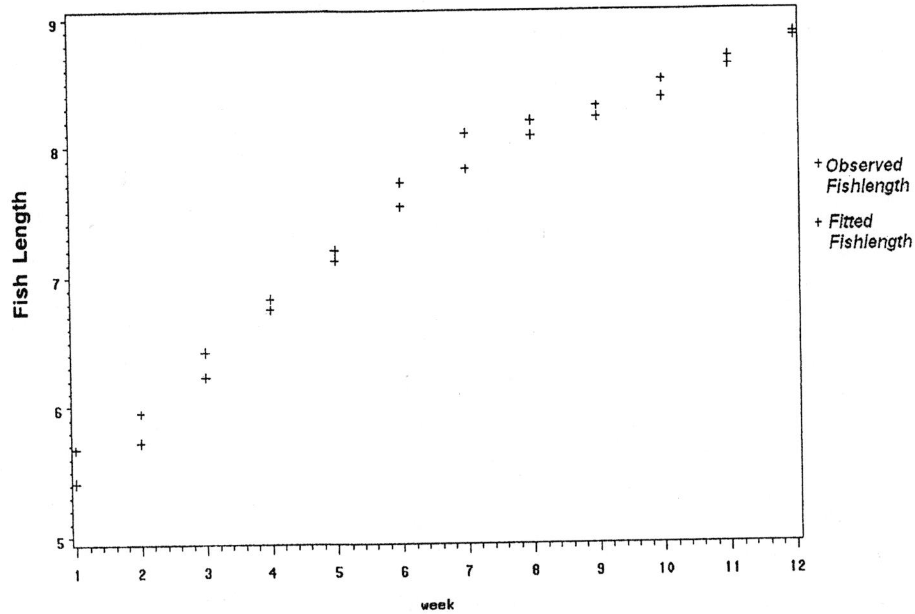

Graph 8.1: VB Fitting of the Fish Length-data

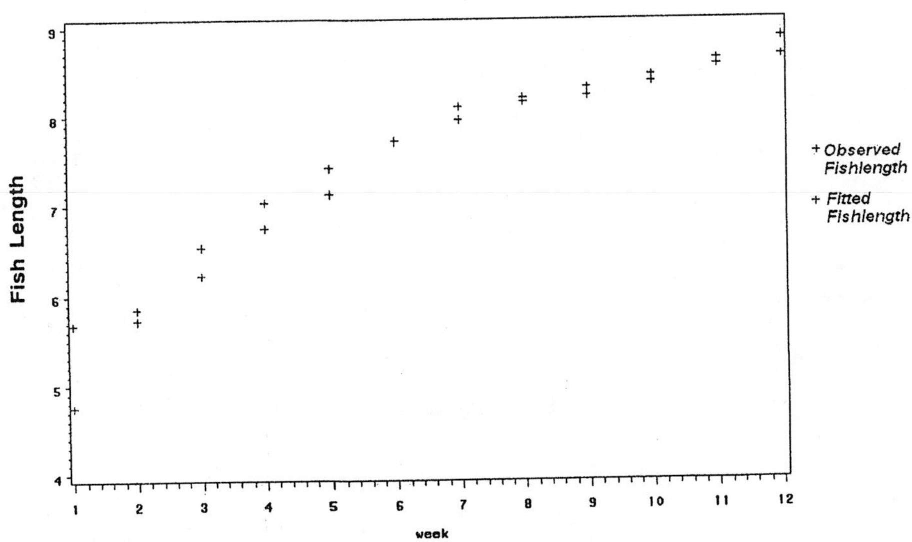

Graph 8.2: Putter Fitting of the Fish Length-data

In order to evaluate the goodness of fit the estimated values are compared against the corresponding observed values and the value of R-square criterion is determined with respect to each model. Though all fits are extremely good (R-square value more

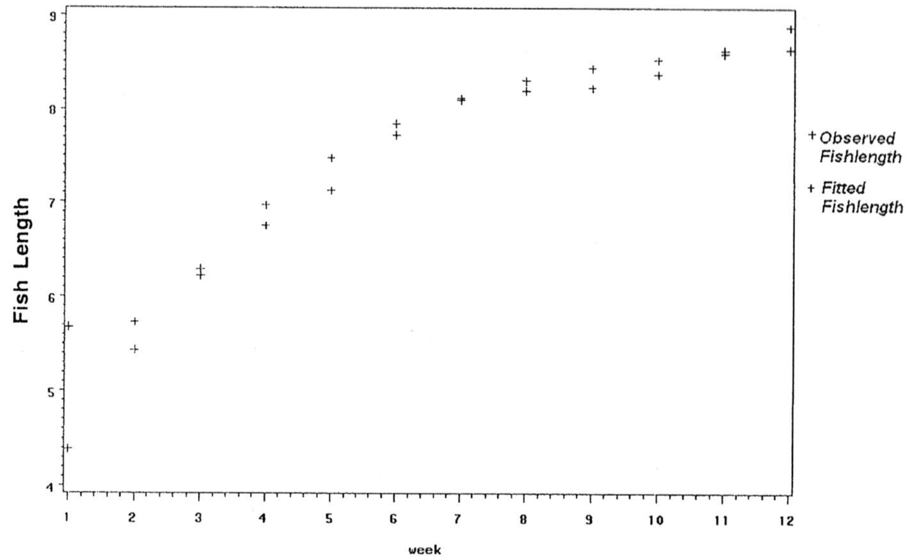

Graph 8.3: Gompertz Fitting of the Fish Length-data

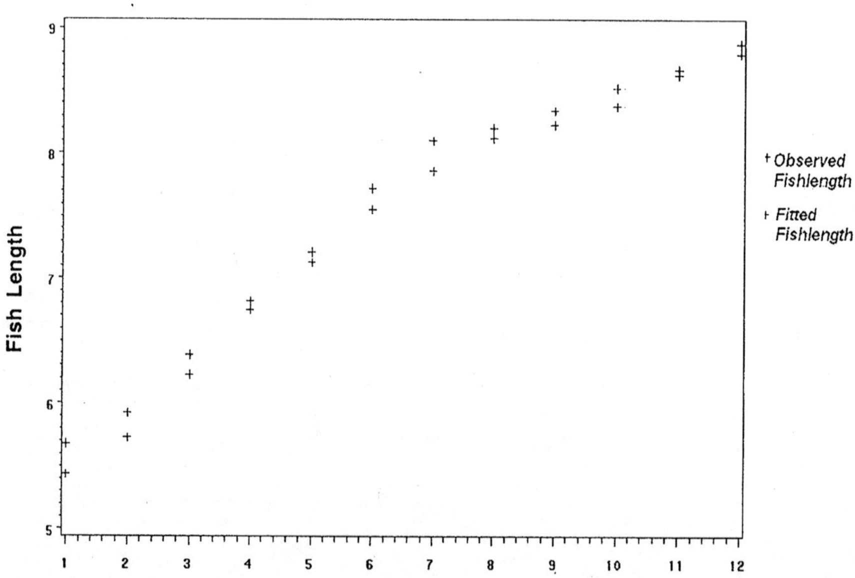

Graph 8.4: Richards Fitting of the Fish Length-data

than 0.9, excepting Gompertz model), Richards model turns out to be the best. VB and Logistic fits follow Richards neck to neck. The fitted and observed values are represented in five graphs addressing the five models.

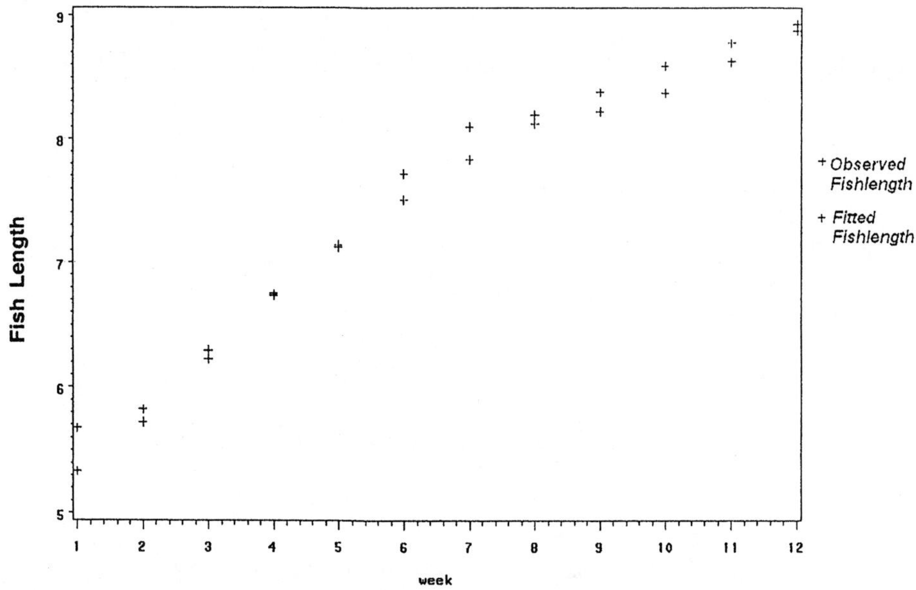

Graph 8.5: Logistic Fitting of the Fish Length-data

Finally, the fitted model equations are obtained as follows:

1. VB model (deterministic): $L_t = 11.00 \times [1\text{-exp}(-0.08637 \times (t+7.18)]$ (6)
2. VB model (non-linear fit): $L_t = 10.095 \times [1\text{-exp}(-0.1205 \times (t+5.3820)]$ (7)
3. Putter model: $L_t^{(p)} = 9.2151 \times [1\text{-exp}(-0.1475 \times t)]$^(1/3) (8)
4. Gompertz model: $L_t^{(g)} = 8.7388 \times \exp(-\exp(-0.3714 \times t)$ (9)
5. Richards model: $L_t^{(R)} = 9.289 \times (1+1.3527 \times \exp(-0.2391 \times t)]$^ (-1/1.3527) (10)
6. Logistic model: $L_t^{(L)} = 9.663 \times [(1+\exp(-0.2084 \times \text{week})]$^(-1)) (11)

Conclusion

The illustrated example of fishery model involving growth and time variables is expected to help future researchers in pursuing research in this area.

Modelling, Forecasting, Artificial Neural Network and Expert System in Fisheries
Editors: Ajit Kumar Roy & Niranjan Sarangi
Published by: DAYA PUBLISHING HOUSE (2009)

Pages 112–123

Chapter 9

Quantitative Techniques for Fish Stock Assessment

R.S. Biradar

Introduction

It was generally believed that the fishery resources are inexhaustible and can be exploited forever without depleting their stocks. Of late, it has been realized by Fishery Managers and Stakeholders that fishery resources are limited and need appropriate management strategies for sustainable fisheries development. Management of capture fisheries requires assessment of fish stocks and determining the optimum level of fishing in order to obtain sustainable fish production over the years. It is in this context that fish stock assessment techniques assume importance in tendering advice on optimum exploitation of fish stocks.

Assessment of fish stocks involves the application of statistical and mathematical calculations to relevant data for obtaining quantitative understanding of reactions of stocks to alternative management practices. The direct impact of fishing on an exploited stock forms the basis for more complex and realistic analysis of stock dynamics. Methodologies of fish stock assessment were initially developed for single species and single gear, which were later generalized for more complex multispecies and multigear systems, which are data and computation intensive. Stock assessment techniques address the issues such as:

(*a*) How big are the resources ?

(*b*) How many fish can be caught each year without adversely affecting the stock ?

(*c*) How many boats (effort) can be employed for harvesting ?

Traditionally fishery scientists have been concerned with the estimation of maximum sustainable yield (MSY) of a fishery. The MSY is the largest catch that can be obtained year after year without causing the collapse of the stock. Fish stock assessment methods are useful in determining MSY of fisheries. In this chapter, some important single species fish stock assessment models are discussed.

Stock Assessment Models

Two types of models are essentially employed in fish stock assessment *viz.* surplus production and analytical models.

Surplus Production Models

These are also called the global models or the macro models or the holistic models. These models deal with the observable inputs (fishing effort) and actual outputs (yield in weight) and do not consider the influence of factors such as recruitment, growth and natural mortality, etc. on the population.

The key process modeled by these models is biomass regeneration. These models deal with changes at population level and do not take into account the age distribution of the stock. The central feature of these models is surplus production.

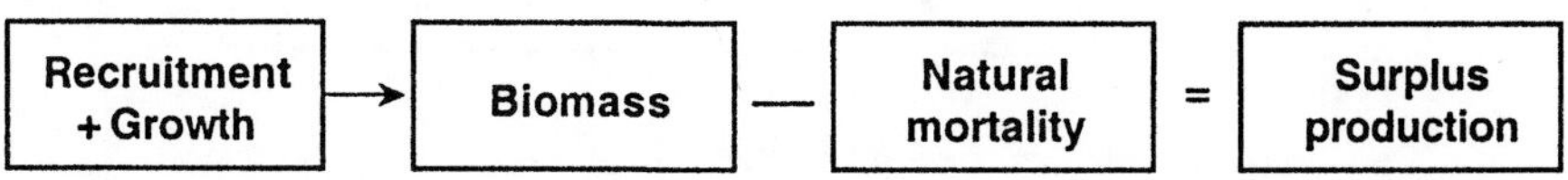

If the yield from a fishery is exactly equal to the surplus produced then the stock is said to be in equilibrium condition. These models establish the relationship between stock size and the surplus produced. Graham (1935) was the first to introduce these models but are often referred to as 'Schaefer models' (Schaefer, 1954). These models are simple and the data requirements are less as compared to analytical models.

The basic assumptions underlying these models are:

(*i*) They deal with unit stock
(Unit stock is a subset of one species having the same growth and mortality parameters and occupying a particular geographical area)

(*ii*) The population is in equilibrium

(*iii*) The population reacts instantaneously to any change in the fishing effort

(*iv*) The environment, the carrying capacity and recruitment are assumed constant.

These models require time series data on catch and effort.

Schaefer Model

Let B denote the current biomass and $B\infty$ the carrying capacity of the aquatic ecosystem. If $B < B\infty$, then the population will increase. The rate of increase will be some function of the current biomass *i.e.,*

$$\frac{dB}{dt} = f(B) \quad (1)$$

As the function f(B) will be zero when B = 0 and when B = B∞, the simplest equation that satisfies these conditions is

$$\frac{dB}{dt} = rB\,(B_{\infty} - B) \tag{2}$$

where, r is the intrinsic rate of growth. If catch Y is taken during the period then total change in the population is

$$\frac{dB}{dt} = rB\,(B_{\infty} - B) - Y \tag{3}$$

By definition, under equilibrium conditions,

$$\frac{dB}{dt} = 0 \text{ and hence}$$

$$Y = rB\,(B\infty - B) \tag{4}$$

Equation (4) is in the form of a parabola (Figure 9.1) and suggests that maximum yield occurs when B = B∞/2.

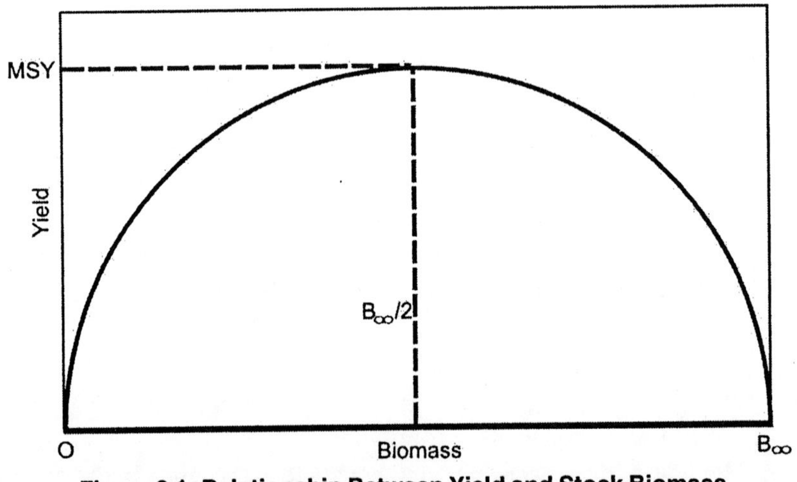

Figure 9.1: Relationship Between Yield and Stock Biomass

Assuming that the catch per unit of effort (U=Y/f) is an index of abundance, Schaefer formulated the model (Equation No.4) in terms of observable catch per unit effort and the corresponding fishing effort as

$$U = a + bf \tag{5}$$

and $$Y = af + bf^2 \tag{6}$$

where, f is fishing effort.

The expected straight-line relationship of catch per unit effort (cpue) with fishing effort is shown in Figure 9.2 and parabolic relationship of yield with effort is given in Figure 9.3.

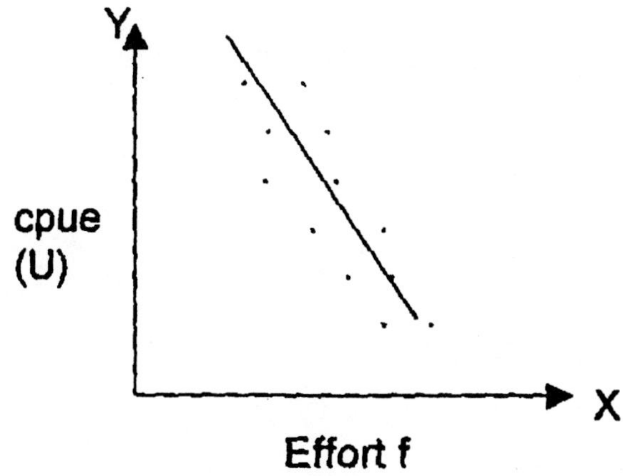

Figure 9.2: Catch Per Unit Effort Versus Effort

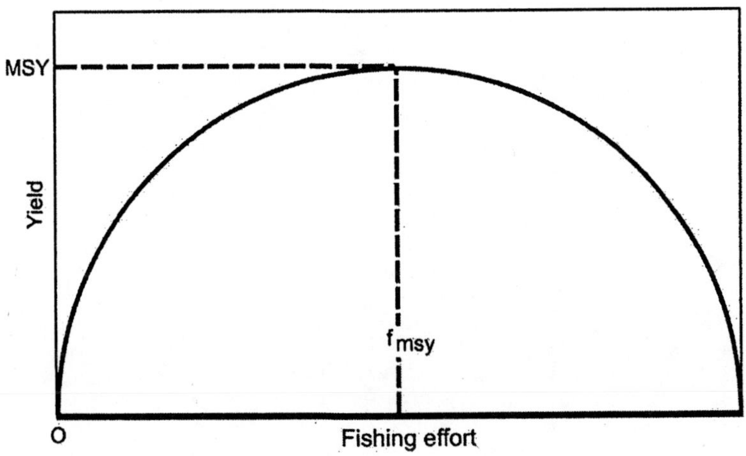

Figure 9.3: Yield Versus Fishing Effort

Equation No. (6) *i.e.* $Y = af + bf^2$ is the basis for obtaining the management reference points and to assess the status of the stock. In the above equations, U is the catch per unit effort, f is the fishing effort, Y is the yield (or catch), a and b are constants to be estimated.

Based on equation No.(6), the f_{msy} and MSY are given by

$$f_{msy} = -a/2b \qquad (7)$$

$$MSY = -a^2/4b \qquad (8)$$

where, f_{msy} is the amount of fishing effort required to obtain the MSY.

Fox Model

Fox (1970) postulated an exponential form of the Schaefer's model that is given by

$$\ln U = c + df$$

i.e., $\ln (Y/f) = c + df$

i.e., $Y/f = \exp (c+df)$ (9)

i.e., $Y = f \exp (c+df)$ (10)

In the above equations, c and d are constants.

Based on equation no. (10), the f_{msy} and MSY are given by

$$f_{msy} = -1/d \quad (11)$$

$$MSY = (-1/d) \exp (c-1) \quad (12)$$

In this model, the cpue (U) is expected to decrease exponentially with the increase in effort. In other words, logarithm of cpue is expected to decrease linearly with respect to effort (Figure 9.4).

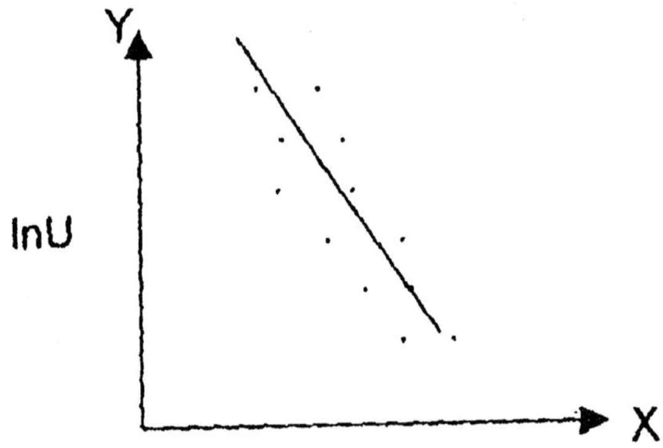

Figure 9.4: Relationship Between Logarithm of Cpue and Effort

Both the models conform to the assumption that the cpue declines as the effort increases. The Schaefer model implies that at high level of effort (f =–a/b), cpue becomes zero. However, Fox model implies non-zero cpue even at higher levels of effort. Thus, for a given series of catch and effort, either of the models depending upon the behaviour of the cpue with the effort can be fitted. It has to be noted that any departure from the assumption of equilibrium conditions will invalidate the models and may result in erroneous estimates of f_{msy} and MSY.

Limitations of Surplus Production Models

1. These models ignore the real biological processes which actually generate the biomass,
2. These models do not consider age structure of the population,

3. Equilibrium conditions assumed by these models are rarely met in any fishery,
4. The concept of MSY is least appropriate for short-lived highly fecund species where yield is directly related to recruitment, which varies from year to year under the influence of fishery independent factors.

Dynamic Pool Models

These models are also called analytical or micro-models. These models explicitly consider the role of the individual parameters such as growth, natural mortality, fishing mortality, age at first capture etc. on the production process in the exploited stocks.

Beverton and Holt Model

One of the best-known and commonly used dynamic pool model was developed by Beverton and Holt (1957). It is a steady state model. When a stock is in steady state, the annual yield from the entire stock is equal to the yield from a single group of recruits during its entire life span. Because of this, it suffices to compute the total yield from a single recruit group over its life span. The annual yield is directly proportional to the level of annual recruitment for any fixed combination of the other factors such as growth and mortality. As the size of the stock and catch change in direct proportion to change in the number of recruits, it is unnecessary to know absolute magnitude of recruitment in order to find the optimum value of age at first capture and fishing mortality. Hence, it is sufficient to find the values that maximize the yield from any arbitrary number of recruits and therefore, the yield per recruit can be worked out. The Beverton and Holt yield per recruit (Y/R) equation makes use of parameters of Von-Bertalanffy growth function (VBGF) besides other fisheries related parameters. The Beverton and Holt yield per recruit (Y/R) equation is given by

$$Y/R = F\exp(-M(t_c - t_r))\, W\infty\, [1/Z - 3S/(Z+K) + 3S^2/(Z+2K) - S^3/(Z+3K)] \quad (13)$$

where,

S: $\text{Exp}[-K(t_c - t_o)]$

$W\infty$: asymptotic weight (a parameter of VBGF)

F: Instantaneous rate of fishing mortality

M: Instantaneous rate of natural mortality

t_c: Age at first capture

t_r: Age at recruitment

Z: F + M

K: Growth coefficient (a parameter of VBGF)

t_o: age at which length is zero (a parameter of VBGF)

It can be seen that Y/R is a function of two variables F and t_c, which are of direct consequence to fisheries management as F is proportional to the effort f and t_c is a function of gear selectivity (mesh size). Different values of Y/R are obtained for different combinations of F and t_c given the values of $W\infty$, M, t_r, t_o and K. The model is useful in estimating (i) the optimum effort level and (ii) optimum mesh size of the gear.

The model requires age or size catch composition data (from aging or length frequency studies).

Y/R is presented in one or more of the following ways:

(*a*) Y/R is depicted as a function of F, keeping t_c constant (Figure 9.5). The shape of Y/R curve is affected by M (Figure 9.5).

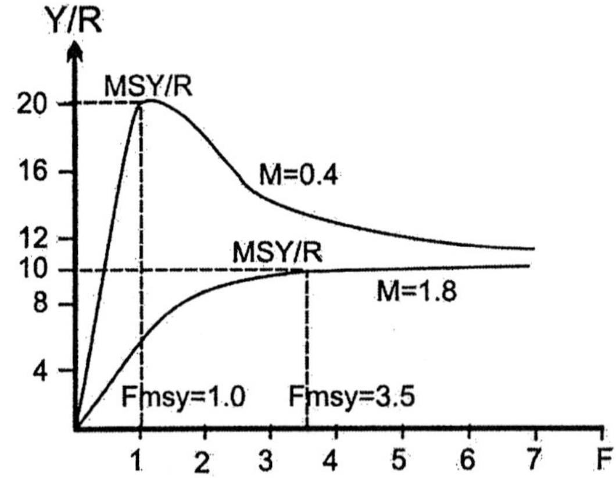

Figure 9.5: Y/R as a Function of F

It is observed from Figure 9.5 that lower M requires a lower F_{msy} and produces a higher MSY/R. Fishing effort above f_{msy} will lead to severe reduction in the catch. If M is high, Y/R curve do not have pronounced maximum and it is difficult to estimate F_{msy} by the Y/R curve.

(*b*) Y/R is depicted as a function of t_c keeping F constant.

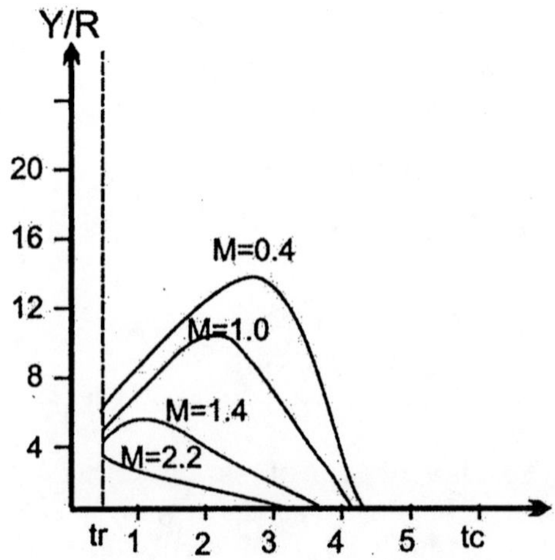

Figure 9.6: Y/R as a Function of t_c

It is seen from Figure 9.6 that Y/R is higher when M is lower. Further, higher Y/R is obtained at lower t_c when M is high.

(*c*) In Figure 9.7, Y/R is depicted as a function of F and t_c (yield isopleth).

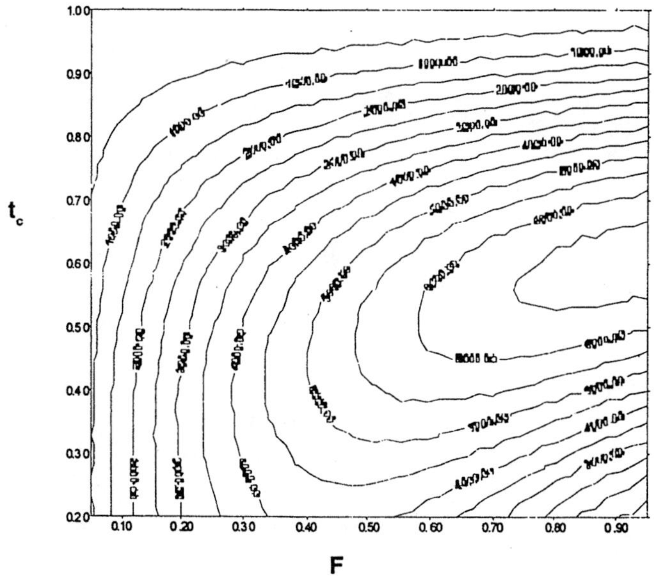

Figure 9.7: Yield Isopleth

Assumptions of Y/R Model

1. The stock is in steady state equilibrium.
2. Recruitment is constant.
3. Knife-edge recruitment and Knife-edge selection are assumed.
4. The fishing and natural mortality are constant from the time they are fully exploited.
5. Only the natural mortality is affecting the stock from the age at recruitment to the age at first capture.
6. Growth in length follows the VBGF.
7. The length weight relationship is isometric (*i.e.*, the exponent b = 3).

Advantages of Y/R Model

1. The model helps in viewing the long-term yields with the simultaneous changes in the fishing effort and the mesh size.
2. The model has more biological realism than surplus production models.
3. Expressing yield as a fraction of recruits is a convenience that removes the need to know the absolute values of recruitment.

Disadvantages

1. The model is applicable only to steady state conditions.
2. Yield is not expressed in absolute terms.
3. For fish stocks with high natural mortality, it is difficult to estimate F_{msy} by Y/R curve.

Beverton and Holt's Biomass Per Recruit Model

For stocks with high M, the Y/R curve does not have a maximum. In such situations, one may come to the wrong conclusion that efforts could be increased indefinitely. In such cases, it is advisable to look at biomass per recruit (B/R) curve. The B/R is related to Y/R as given below:

$$\overline{B}/R = (1/F)\,(Y/R)$$

where, $\overline{B}$ is the average biomass during a year.

It is always recommended to depict both Y/R and $\overline{B}$/R curves, especially for stocks with high M, as the latter curve gives what percentage of biomass corresponds to the optimum F level F_{msy}. It is always advisable to maintain certain minimum level of biomass to avoid recruitment over fishing. Figure 9.8 depicts Y/R and $\overline{B}$/R curves.

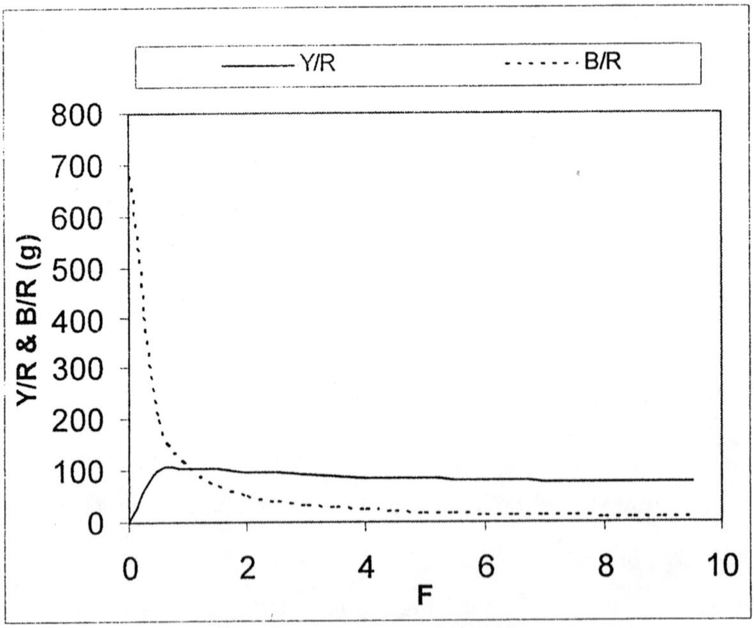

Figure 9.8: Y/R and B/R Curves

Virtual Population Analysis (VPA)

Both virtual population analysis (VPA) and cohort analysis which is an approximation of VPA use the number of fish caught in different age classes during commercial fishing operations to estimate stock size and fishing mortality for each age class. These methods if applied to time series data are thus very useful in providing the inputs required for estimating stock-recruitment relationship. The knowledge on stock-recruitment relationship enables computation of absolute yield and biomass of the stock. Cohort analysis has been extended to deal with situations when only length composition data are available (Jones, 1984), as encountered in tropical fisheries.

Models for Data Scarce Situations

In case of poorly investigated stocks, time series data of catch and effort or age or size composition data may not be available. However, some estimates of over all biomass and natural mortality may be available. In such situations, some empirical formula has been developed with the objective of providing a rough estimate of the MSY based on such scanty data. Formula developed by Gulland (1971), which was later modified by Cadima (in Troadec, 1977) are discussed here.

Gulland's Formula

Gulland (1971) suggested the following formula to estimate MSY:

$$MSY = 0.5\ MBv$$

where, M is natural mortality and Bv is virgin stock biomass.

Cadima's Formula

The Gulland (1971) formula is applicable for virgin stocks. In the case of exploited fish stocks for which only limited stock assessment data are available, Cadima (in Troadec, 1977) suggested the following formula:

$$MSY = 0.5\ Z\ \bar{B}$$
$$= 0.5\ (Y + M\bar{B})$$

where, Y is the total catch, $\bar{B}$ is the average (annual) biomass and Z is the total mortality.

The biomass Bv or $\bar{B}$ in the above formula are estimated by exploratory bottom trawl surveys and/or acoustic surveys.

Multi-species/Multifleet Models

The models and methods described so far are applicable mostly to a single stock exploited by single fleet. In a situation where single fleet exploits several stocks and several fleets compete for exploiting the same resource, the biological, economic and technical interactions come into play. Sparre and Venema (1992) suggest that the simplest way to deal with multispecies/multifleet system is to apply surplus production models to the total catch of all species and the total effort by all fleets. General applicability of this approach can be questioned as it combines so many complex interactions. Multispecies Virtual Population Analysis has been suggested to take care of biological interactions between species. The data requirement of these models is very high. BEAM4, Multispecies, multigear Y/R model development by Sparre and Willmann (1992) is useful in analysis of technical interactions between fishing gears. FiSAT has the provision to apply BEAM4 model upto 19 species and 12 fishing gears (fleets) under Thompson and Bell yield prediction model. Although the data requirements are high, the model may be used to investigate the effect of mesh sizes, gear bans and closed seasons as well as changes in fishing mortality rates (Hoggarth *et al.*, 2005).

Computer Software Packages for Fish Stock Assessment

Number of computer software packages is available for fish stock assessment. FAO (Sparre, 1987) developed the length based fish stock assessment (LFSA) package to cater to the needs of tropical fish stock assessment. Complete ELEFAN package was later developed by ICLARM (Gayanilo, Soriano and Pauly, 1988). Integrating the best features of both the packages, a new package entitled 'FAO-ICLARM Stock Assessment Tools' or FiSAT was developed later in 1993. The FiSAT package estimates growth parameters of VBGF namely $L\infty$, K, t_0, mortality parameters Z, M and F and gear selectivity parameters. Virtual population analysis (VPA), yield per recruit, Thompson and Bell single/multispecies, multigear yield and stock prediction analysis can also be carried out with FiSAT package.

Fisheries Management Science Programme (FMSP) of UK Department for International Development (DFID) has developed four software tools *viz.*: Length Frequency Distribution Analysis (LFDA), Yield, Catch Effort Data Analysis (CEDA) and Participatory Fisheries Stock Assessment (Par Fish). The LFDA package estimates VBGF growth parameters $L\infty$, K and t_0 and total mortality rate Z, M, MSY and F_{msy}. The yield package estimates among other parameters Y/R and B/R, MSY, f_{msy}, F_{msy} and B_{msy} allowing for uncertainty in input parameters. The CEDA package estimates the current and unexploited fish population size, the catchability coefficient and other intermediate parameters, based on long time series of catch and effort data. Provision is also available for fitting production models under non-equilibrium conditions. The Par Fish software package can also be used to fit the 'Schaefer' biomass dynamic model used in CEDA. With its Bayesian approach, the CEDA software can be used when little or no catch/effort data are available and in a co-management situation. For further details on FMSP tools, the readers may refer to Hoggarth *et al.* (2005) and Hoggarth *et al.* (2006).

References

Beverton, R. J. H. and Holt, S. J., 1957. On the dynamics of exploited fish populations. *Fish. Invest. Minist. Agric. Fish. Food. G.B.* (2 sea Fish.), 19:533p.

Fox, W. W. Jr., 1970. An exponential surplus-yield model for optimizing exploited fish populations. *Trans. Am. Fish. Soc.*, 99:80-88.

Gayanilo Jr., F.C., Soriano, M. and Pauly, D., 1988. A draft guide to the COMPLEAT ELEFAN, *ICLARM Software Project* 2:65p.

Graham, M., 1935. Modern theory of exploiting a fishery and application to North sea trawling. *J. Cons. CIEM*, 10(3):264-74.

Gulland, J. A. (comp.), 1971. The fish resources of the ocean. West Byfleet, Surrey, Fishing News (Books), Ltd., for FAO, 255p. Revised edition of FAO Fish. *Tech. Pap.*,97:425p.(1970).

Hoggarth, D. D., Abeyasekere, S., Arthur, R.I., Beddington, J.R., Burn, R.W., Halls, A.S., Kirkwood, G.P., Allister, Mc, Medley, P., Mees, C.C., Parkes, G.B., Pilling, G.M., Wakeford, R.C. and Welcomme, R.L., 2006. Stock assessment for fishery management. *FAO Fish. Tech. Pap.*, 487 : 261p.

Hoggarth, D. D., Mees, C. C., O'Neill, C., Hindson, J. and Krishna, M., 2005. A guide to Fisheries Stock Assessment using the FMSP Tools. Marine Resources Assessment Group (MRAG), London, 91p.

Jones, R., 1984. Assessing the effects of changes in exploitation pattern using length composition data (with notes on VPA and Cohort analysis). *FAO Fish. Tech. Pap.*, 256: 118p.

Schaefer, M., 1954. Some aspects of the dynamics of populations important to the management of the commercial marine fisheries. *Bull.I-ATTC/Bol.CIAT*, 1(2):27-56.

Sparre, P. and Venema, S. C., 1992. Introduction to tropical fish stock assessment. Part 1, Manual (Rev.2), *FAO Fish. Tech. Pap.* (306.1):407p.

Sparre, P. and Willmann, R., 1992. Computer programs for bio-economic analysis of fisheries. BEAM4 Manual Analytical bio-economic simulation of space-structured multi-species and multi-fleet fisheries. FAO Computerized Information Series (Fisheries), No.3.

Sparre, P., 1987. Computer programs for fish stock assessment. Length based fish stock assessment for Apple II computers. *FAO Fish. Tech. Pap.* (101), Suppl.2:218p.

Troadec, J. P., 1977. Methods semi-quantitative d'evaluation. *FAO Circ. Perches*, 701:131-141.

Modelling, Forecasting, Artificial Neural Network and Expert System in Fisheries
Editors: Ajit Kumar Roy & Niranjan Sarangi
Published by: DAYA PUBLISHING HOUSE (2009)
Pages 124–147

Chapter 10

Time Series Modelling and its Applications in Fisheries

S. Ravichandran

Introduction

Time series data on various aspects of fish *viz.* landing, catch, effort, export of marine products etc are collected over time. Box-Jenkins Auto Regressive Integrated Moving Average (ARIMA) time series methodology is widely used for modelling these time series data (Box *et al.*, 1994). This methodology can be applied only when either the series under consideration is stationary or it can be made so by differencing, de-trending, or by any other means. Another disadvantage is that this approach is empirical in nature and does not provide any insight into the underlying mechanism. An alternative mechanistic approach, which is quite promising, is the "Structural time series modelling (Harvey, 1996)". Here, the basic philosophy is that characteristics of the data dictate the particular type of model to be adopted from the family. In this chapter, we discuss about ARIMA and STM methodologies utilised for modelling time series data in the presence of trend, seasonal, cyclical fluctuations, and autoregressive processes with the help of suitable illustrations from Indian fisheries. Details of software employed for fitting of ARIMA and STM models is discussed.

ARIMA Modelling

ARIMA method consists in developing a suitable ARIMA model, which in turn provides a comprehensive tool kit for time series model identification, parameter estimation and forecasting. Box–Jenkins time series models amalgamate three types of processes, *viz.* autoregressive (AR) of order p, differencing to make a series of degree d and moving average (MA) process of order q and hence is written as ARIMA (p, d, q). This process assumes that the series under consideration is stationary. Main aim of analyzing any time series data is to forecast future values of the series under study. ARIMA model approach deals with short term forecasting rather than long term forecasting.

Structural Time Series Modelling (STM)

Structural time series models are formulated in such a way that their components are stochastic; in other words, they are regarded as being driven by random disturbances. The key to handling structural time series models is the state space form, with the state of the system representing the various unobserved components such as trends, cycles and seasonals. Once in state space form (SSF), the Kalman filter provides the means of updating the state, as new observations become available. Predictions are made by extrapolating these components into the future. Kalman filter, a smoothing algorithm is used for obtaining the best estimate of the state at any point within the sample. Prediction and smoothing is carried out once the parameters governing the stochastic movements of the state variables have been estimated. Estimation of these parameters, known as "hyper parameters", is based on Kalman filter. Once a model is estimated, its suitability can be assessed using goodness of fit statistics.

Basic Structural Time Series Model (BSM)

A "Structural time series model" is set up in terms of its various components, like trend, cyclical fluctuations, and seasonal variations, *i.e.*

$$Y_t = T_t + C_t + S_t + \varepsilon_t, \tag{1}$$

where, Y_t is the observed time series at time t, T_t, C_t, S_t, ε_t are the trend, cyclical, seasonal and irregular components.

Local Level Model (LLM)

In the absence of seasonal and cyclical components, eq. (1) reduces to

$$Y_t = \mu_t + \varepsilon_t , \quad \varepsilon_t \sim N(0, \sigma_\varepsilon^2) , \quad t = 1, 2, \ldots, T . \tag{2}$$

When the trend component (μ_t) does not show a steady upward or downward movement, it becomes a permanent component called "level". Sometimes, it is assumed to vary according to a random walk, *i.e.*

$$\mu_t = \mu_{t-1} + \eta_t , \quad \eta_t \sim N(0, \sigma_\eta^2) . \tag{3}$$

Eqs. (2) and (3) together form the LLM. It may be noted that these equations are already in state space form. However, level μ_t is not directly observable but can be estimated. Further, forecast values are weighted average of the data points. When $\sigma_\varepsilon^2 = 0$, forecast is just the last observation and when $\sigma_\eta^2 = 0$, level is constant and the best forecast is sample mean. Level of time series vary over time depending on signal to noise ratio $q = \sigma^2_\eta / \sigma^2_\varepsilon$. Estimation of μ_t, conditional on σ^2_ε and σ^2_η, is done recursively using Kalman filter and smoother (Harvey, 1996). The parameters σ^2_ε and σ^2_η are unknown and are treated as hyperparameters. Likelihood function can be evaluated by Kalman filtering via prediction error decomposition (Shumway and Stoffer, 2000). Once σ^2_ε and σ^2_η are known, one-step-ahead prediction of level, *i.e.* estimator of μ_{t+1} given $Y_t = \{Y_1, Y_2, \ldots, Y_t\}$, *viz.*

$$a_{t+1} = E(\mu_{t+1} \mid Y_t), \tag{4}$$

is evaluated recursively by Kalman filter. Prediction error variance

$$P_{t+1} = Var(\mu_{t+1} \mid Y_t) = Var(a_{t+1}) \tag{5}$$

is also obtained recursively. Reduced form of LLM is ARIMA (0, 1, 1) model.

Local Linear Trend Model (LLTM)

LLTM is given by eq. (2) along with the following two equations:

$$\mu_t = \mu_{t-1} + \beta_{t-1} + \eta_t, \tag{6}$$

$$\beta_t = \beta_{t-1} + \xi_t, \quad t = \ldots, -1, 0, 1, \ldots, \tag{7}$$

where, $\eta_t \sim N(0, \sigma^2_\eta)$ and $\xi_t \sim N(0, \sigma^2_\xi)$. It may be mentioned that η_t, ξ_t and ε_t are independent of one another. If $\sigma^2_\eta = \sigma^2_\xi = 0$, eqs. (6) and (7) collapse to

$$\mu_t = \mu_{t-1} + \beta, t = 1, 2, \ldots, T, \tag{8}$$

which can equivalently be written as

$$\mu_t = \alpha + \beta t, t = 1, 2, \ldots, T, \tag{9}$$

showing that the deterministic linear trend is a limiting case.

LLTM is in state space form with state vector $\alpha_t = (\mu_t, \beta_t)$. Updating and prediction are carried out using Kalman filter by assuming that σ^2_ε, σ^2_η and σ^2_ξ are known. Otherwise these can be estimated using maximum likelihood method for state space models (de Jong, 1988; Koopman and Shephard, 1992). Reduced form of a LLTM is ARIMA (0, 2, 2) model. Andrews (1994) showed that forecast function of LLTM performs better than that of corresponding ARIMA model.

Local Linear Trend Model with Intervention Effect (LLTMI)

Intervention analysis is concerned with making inference about effects of known events. These effects are measured by including intervention, or dummy variables in a dynamic model (Harvey and Durbin, 1986). LLTMI is described by the following equations:

$$Y_t = \mu_t + \lambda w_t + \varepsilon_t \tag{10}$$

$$\mu_t = \mu_{t-1} + \beta_{t-1} + \lambda w_t + \eta_t \tag{11}$$

$$\beta_t = \beta_{t-1} + \lambda w_t + \xi_t, \tag{12}$$

where, w_t is intervention variable and λ is its coefficient. The quantity w_t depends on the form which intervention is assumed to take. As for LLTM, estimation of state vector $\alpha_t = (\mu_t, \beta_t)$ for LLTMI is similarly carried out by putting the model in state space form and applying Kalman filter recursively by treating w_t as an explanatory variable.

Goodness of FIT

Goodness of fit statistics is used for assessing overall model fit. Basic measure of goodness of fit in time series models is prediction error variance. Comparison of fit between different models is based on Akaike information criterion (AIC) and Schwartz–Bayesian information criterion (SBC) where

$$AIC = -2 \log L + 2n \text{ and } SBC = -2 \log L + n \log T,$$

where, L is the likelihood function, which is expressed in terms of estimated one–step–ahead prediction errors $\hat{v}_t = Y_t - \hat{Y}_{t|t-1}$. Here n is the number of hyper parameters estimated from the model and T is the total number of observations. Lower the values of these statistics, better is the fitted model.

Software for STM Modelling

STM models can be fitted to the data using Structural Time Series Analyser, Modeller and Predictor (STAMP) software package (Koopman *et al.*, 2000) or by using SsfPack2.2 (Koopman *et al.*, 1999) software package or by SAS (Statistical Analysis System) Proc UCM. SAS version 9.0 includes an experimental version of Proc UCM, Selukar (2006). Proc. Statepace can be used in earlier version for this kind of analysis. SsfPack is also available for S-Plus users. STAMP handles a wide variety of models, including basic GARCH and stochastic volatility models. STAMP algorithms are fast, and STAMP SsfPack contains a host of useful utilities for performing analysis with state space models.

Illustration

As an illustration, data on all-India marine, inland, and total fish production during 1980-81 to 2005-06 is utilised (Table 10.1). Using, STAMP statistical software

LLTM is developed for the above data sets. Parameter estimates as well as goodness of fit statistics are presented in Table 10.2. Next, ARIMA (0, 2, 2) model is fitted to the data sets using Statistical Analysis System (SAS) statistical software package. Estimates of constant, MA1, MA2 parameters and goodness of fit statistics are presented in Table 10.3. Evidently, a comparison of goodness of fit statistics indicates that LLTM performs much better than ARIMA (0, 2, 2) model for all data sets and hence forecast for next five years using LLTM is carried out and is presented.

Table 10.1: Fish Production (in thousand tonnes) of India During 1980–81 to 2005-06

Year	*Marine*	*Inland*	*Total*
1980–81	155.5	88.7	244.2
1981–82	144.5	99.9	244.4
1982–83	142.7	94.0	236.7
1983–84	151.7	98.7	250.6
1984–85	169.8	110.3	280.1
1985–86	171.6	116.0	287.6
1986–87	171.3	122.9	294.2
1987–88	165.8	130.1	295.9
1988–89	181.7	133.5	315.2
1989–90	227.5	140.2	367.7
1990–91	230.0	153.6	383.6
1991–92	244.7	171.0	415.7
1992–93	257.6	178.9	436.5
1993–94	264.9	199.5	464.4
1994–95	269.2	209.7	478.9
1995–96	270.7	224.2	494.9
1996–97	296.7	238.1	534.8
1997–98	295.0	243.8	538.8
1998–99	269.6	256.6	526.2
1999–00	285.2	282.3	567.5
2000–01	281.1	284.5	565.6
2001–02	283.0	312.6	595.6
2002–03	299.0	321	620.0
2003–04	294.1	345.8	639.9
2004–05	277.8	352.6	630.4

Table 10.2: Summary Statistics for Local Linear Trend Model

Parameter	*Marine*	*Inland*	*Total*
μ_t	277.80	236.28	639.73
β_t	3.86	13.06	18.33
σ_ε^2	0	10.85	24.17
σ_η^2	208.47	16.55	208.40
σ_ξ^2	0.99	2.13	2.16

Goodness of Fit Statistics

AIC	172.41	137.14	169.65
SBC	176.85	141.37	174.47
Standard Error	14.34	5.86	16.06

One-step-ahead Prediction for 2005-06 (in thousand tonnes)

	282	279.38	658.01
	(277.8)	(352.6)	(630.4)

Note: Figures in parentheses indicate actual values.

Forecast Values (in thousand tonnes)

2006-2007	286.00	281.43	676.34
2007-2008	290.00	300.73	694.67
2008-2009	294.00	326.57	713.00
2009-2010	297.00	334.27	731.32
2010-2011	301.00	333.54	749.65

Table 10.3: Summary Statistics for ARIMA (0, 2, 2) Model

Parameter	*Marine*	*Inland*	*Total*
Constant	20.28	19.87	32.57
MA 1	-0.85	-0.88	–0.67
MA 2	0.22	0.13	-0.34

Goodness of Fit Statistics

AIC	288.32	208.37	292.35
SBC	295.67	215.85	298.35
Standard Error	22.83	9.27	32.45

Since "Structural time-series modelling" approach performs better than ARIMA time–series approach is high time that fishery scientists should start using this modelling methodology.

Structural Time Series Model for Cyclical Fluctations

In the absence of seasonal components, STM model reduces to

$$Y_t = \mu_t + \psi_t + \varepsilon_t. \tag{13}$$

For modelling cyclical fluctuations, the following three models are discussed:

Cycle Plus Noise Model (CNM)

Here the trend μ_t is assumed to be constant. Thus, the functional form of CNM (Harvey, 1996) is

$$Y_t = \mu + \psi_t + \varepsilon_t,\ t = 1, 2, \ldots, T. \tag{14}$$

The cyclical fluctuations ψ_t can be expressed as a mixture of sine and cosine terms:

$$\psi_t = \alpha \cos(\lambda_c t) + \beta \sin(\lambda_c t), \tag{15}$$

where, λ_c, $(\alpha^2 + \beta^2)^{1/2}$, $\tan^{-1}(\beta/\alpha)$ represent respectively the frequency, amplitude, and phase. The cycles in eq. (15) need to be made stochastic by allowing the parameters α and β to evolve over time. Following Harvey (1996), the final form of eq. (4) can be written as

$$\begin{bmatrix} \psi_t \\ \psi_t^* \end{bmatrix} = \rho \begin{bmatrix} \cos\lambda_c & \sin\lambda_c \\ -\sin\lambda_c & \cos\lambda_c \end{bmatrix} \begin{bmatrix} \psi_{t-1} \\ \psi_{t-1}^* \end{bmatrix} + \begin{bmatrix} k_t \\ k_t^* \end{bmatrix}, \tag{16}$$

where, the correlation coefficient $\rho \in [0, 1]$ is a damping factor, and k_t and k_t^* are uncorrelated white-noise disturbance terms. Further, $\psi_0 = \alpha$, $\psi_0^* = \beta$. The new parameters are ψ_t, the current value of the cycle, and ψ_t^*, which appears by construction in order to form ψ_t. Eq. (16) is a vector AR (1) process.

Let L denote the lag operator, *i.e.*

$$L\psi_t = \psi_{t-1}. \tag{17}$$

Then eq. (15) can be written as

$$\begin{bmatrix} 1 - L\rho\cos\lambda_c & -L\rho\sin\lambda_c \\ L\rho\sin\lambda_c & 1 - L\rho\cos\lambda_c \end{bmatrix} \begin{bmatrix} \psi_t \\ \psi_t^* \end{bmatrix} = \begin{bmatrix} k_t \\ k_t^* \end{bmatrix} \tag{18}$$

Substituting for ψ_t in eq. (3) yields

$$Y_t = \mu + \frac{(1 - L\rho\cos\lambda_c)k_t + (L\rho\sin\lambda_c)k_t^*}{(1 - 2L\rho\cos\lambda_c + L^2\rho^2)} + \varepsilon_t, \quad t = 1, 2, \ldots, T, \tag{19}$$

where, ε_t is assumed to be uncorrelated with k_t and k_t^*.

For estimation of parameters, eq. (19) has to be put in state space form (Meinhold and Singpurwalla, 1983) and then Kalman filter, prediction and smoothing (Koopman *et al.*, 1999) are applied. In eq. (19), unobservable state, conditional on variances σ_ε^2 and σ_k^2, is done recursively using Kalman filter and smoother. In general, these are unknown and are treated as hyperparameters. Likelihood function can be evaluated by Kalman filter via prediction error decomposition (Shumway and Stoffer, 2000). Maximizing likelihood function with respect to hyper parameters, using quasi-Newton optimization procedure, is referred to as "hyper parameter estimation". After estimation of parameters, prediction and smoothing are performed. The reduced form from ARIMA family corresponding to CNM is "Constant + ARIMA (2,2)".

Trend Plus Cycle Model (TCM)

TCM is given by the following equations:

$$Y_t = \mu_t + \psi_t + \varepsilon_t, \quad t = 1, 2, \ldots, T, \tag{20}$$

$$\mu_t = \mu_{t-1} + \beta_{t-1} + \eta_t, \tag{21}$$

$$\beta_t = \beta_{t-1} + \xi_t, \quad t = \ldots, -1, 0, 1, \ldots \tag{22}$$

In this, ε_t, η_t and ξ_t are the disturbance terms which follow Gaussian distributions with means 0 and variances σ_ε^2, σ_η^2 and σ_ξ^2, called hyper parameters of the model. As mentioned in CNM model, estimation of state vector and hyper parameters is carried out by putting the model in state space form and subsequently Kalman filter is applied with proper initial values. The state space formulation of TCM is as follows:

The measurement equation is

$$Y_t = \begin{bmatrix}1 & 0 & 1 & 0\end{bmatrix}\alpha_t + \varepsilon_t, \quad t = 1, 2, \ldots, T, \tag{23}$$

where as the transition equation is

$$\alpha_t = \begin{bmatrix} \mu_t \\ \beta_t \\ \cdots \\ \psi_t \\ \psi_t^* \end{bmatrix} = \begin{bmatrix} 1 & 1 & . & & \\ 0 & 1 & . & & \\ \cdots & \cdots & \cdots & \cdots & \cdots \\ & & . & \rho\cos\lambda_c & \rho\sin\lambda_c \\ & & . & -\rho\sin\lambda_c & \rho\cos\lambda_c \end{bmatrix} \begin{bmatrix} \mu_{t-1} \\ \beta_{t-1} \\ \cdots \\ \psi_{t-1} \\ \psi_{t-1}^* \end{bmatrix} + \begin{bmatrix} \eta_t \\ \xi_t \\ \cdots \\ k_t \\ k_t^* \end{bmatrix} \qquad (24)$$

The covariance matrix of the vector of disturbances in eq. (24) is a diagonal matrix with diagonal elements { σ_η^2, σ_ξ^2, σ_k^2, $\sigma_{k^*}^2$ }. Once the parameters are estimated using prediction error decomposition, Kalman filter, prediction and smoothing can be applied.

Cyclical Trend Model (CTM)

Here the cycle is actually incorporated within trend. Thus the model is given by the equations (Harvey, 1996):

$$Y_t = \mu_t + \varepsilon_t \, , \quad t = 1, 2, \ldots, T, \qquad (25)$$

$$\mu_t = \mu_{t-1} + \psi_{t-1} + \beta_{t-1} + \eta_t \, , \qquad (26)$$

$$\beta_t = \beta_{t-1} + \xi_t \, . \qquad (27)$$

Estimation of parameters and hyper parameters is carried out by putting the model in state space form and applying Kalman filter. The essential difference between TCM and CTM is that, in the former, the observation Y_t depends on cyclical fluctuations ψ_t explicitly whereas, in the latter, it does so on ψ_{t-1} implicitly through the trend μ_t. ARIMA (2, 2, 4) model is the corresponding analogue of both TCM and CTM.

This particular model can be utilised for modelling time series data having cyclical fluctuations.

STM Models for Describing Season Fluctuations

For modelling seasonal fluctuations in fish populations, A STM is set up in terms of various components of interest, such as trend (μ_t), cyclical fluctuations (ψ_t), seasonal variations (g_t) and error terms (ε_t), *i.e.*

$$Y_t = \mu_t + \psi_t + \gamma_t + \varepsilon_t . \qquad (28)$$

However, if cyclical fluctuations are not present, eq. (19) reduces to

$$Y_t = \mu_t + \gamma_t + \varepsilon_t . \qquad (29)$$

In reality, μ_t varies over time and it seems logical that μ_t depends on μ_{t-1}. We, therefore, let

$$\mu_t = \mu_{t-1} + \beta_{t-1} + \eta_t, \tag{30}$$

$$\beta_t = \beta_{t-1} + \xi_t, \tag{31}$$

where, the error terms ε_t, h_t, x_t are assumed to be mutually uncorrelated and follow respectively $N(0, \sigma_\varepsilon^2)$, $N(0, \sigma_\eta^2)$, and $N(0, \sigma_\xi^2)$ distributions.

Eqs. (29), (30) and (31) together are called "Basic Structural Model" (BSM). The seasonal component γ_t in eq. (29) can be considered as either "Dummy seasonal" or "Trigonometric seasonal" (Harvey, 1996) as described below:

BSM with dummy seasonal (BSMDS)

In the dummy variable form of seasonality

$$\gamma_t = -\sum_{j=1}^{s-1} \gamma_{t-j} + w_t, \tag{32}$$

where, w_t is a white noise disturbance with mean zero and variance σ_w^2. For estimation of parameters, eqs. (29) to (31) have to be first put in state space form (Meinhold and Singpurwalla, 1983). For s = 4, the state space representation is:

$$Y_t = [1 \quad 0 \quad 1 \quad 0 \quad 0]\,\alpha_t + \varepsilon_t, \tag{33}$$

$$\alpha_t = \begin{bmatrix} \mu_t \\ \beta_t \\ \cdots \\ \gamma_t \\ \gamma_{t-1} \\ \gamma_{t-2} \end{bmatrix} = \begin{bmatrix} 1 & 1 & . & & & \\ 0 & 1 & . & & & \\ \cdots & \cdots & . & \cdots & \cdots & \cdots \\ & & . & -1 & -1 & -1 \\ & & . & 1 & 0 & 0 \\ & & . & 0 & 1 & 0 \end{bmatrix} \begin{bmatrix} \mu_{t-1} \\ \beta_{t-1} \\ \cdots \\ \gamma_{t-1} \\ \gamma_{t-2} \\ \gamma_{t-3} \end{bmatrix} + \begin{bmatrix} \eta_t \\ \xi_t \\ \cdots \\ w_t \\ 0 \\ 0 \end{bmatrix} \tag{34}$$

Eq. (33) is called the "observation equation" whereas eq. (34) is the "transition equation". Here α_t is the state vector. The initial state vector, α_0, has a mean of a_0 and covariance matrix P_0. The BSMDS is now in state space form and so Kalman filter is performed to obtain updating, prediction and smoothing of state vector. Prediction and smoothing are carried out only when the hyper parameters, *viz.* σ_ε^2, σ_η^2 and σ_ξ^2 are estimated. This is done using EM algorithm or Newton–Raphson optimization routine (Shumway and Stoffer, 2000).

BSM with Trigonometric Seasonality (BSMTS)

In eq. (29), if γ_t is "trigonometric seasonal" with frequencies $\lambda_j = 2\,p\,j/s, j = 1, 2, \ldots, [s/2]$, the seasonal effect at time t is given by

$$\gamma_t = \sum_{j=1}^{[s/2]} \{\gamma_j \cos(\lambda_j t) + \gamma_j^* \sin(\lambda_j t)\}, \qquad (35)$$

where, [s] denotes integral part of s. When s is even, sine term in eq. (35) disappears for $j = s/2$ and so the number of trigonometric parameters, γ_j and γ_j^*, is always s-1, which is equivalent to the number of coefficients in the "seasonal dummy" formulation. Further, if the trigonometric seasonal components of eq. (35) are incorporated into a linear trend plus error model and the full set of trigonometric terms is included, the two models, *viz.* BSMDS and BSMTS will provide identical results (Harvey, 1996).

For quarterly data, the state space representation of BSMTS is

$$Y_t = \begin{bmatrix} 1 & 0 & 1 & 0 & 1 \end{bmatrix} \alpha_t + \varepsilon_t, \qquad (36)$$

$$\alpha_t = \begin{bmatrix} \mu_t \\ \beta_t \\ \cdots \\ \gamma_{1t} \\ \gamma_{1t}^* \\ \cdots \\ \gamma_{2t} \end{bmatrix} = \begin{bmatrix} 1 & 1 & . & & & . & & & \\ 0 & 1 & . & & & . & & & \\ . & . & . & . & . & . & . & . & . \\ & & . & & & . & & & \\ & & . & & & . & & & \\ . & . & . & . & . & . & . & . & . \\ & & . & & & . & & & \end{bmatrix} \begin{bmatrix} \mu_{t-1} \\ \beta_{t-1} \\ \cdots \\ \gamma_{1,t-1} \\ \gamma_{1,t-1}^* \\ \cdots \\ \gamma_{2,t-1} \end{bmatrix} + \begin{bmatrix} \eta_t \\ \xi_t \\ \cdots \\ w_{1t} \\ w_{1t}^* \\ \cdots \\ w_{2t} \end{bmatrix} \qquad (37)$$

where,

$$C_1 = \begin{bmatrix} \cos(\pi/2) & \sin(\pi/2) \\ -\sin(\pi/2) & \cos(\pi/2) \end{bmatrix} = \begin{bmatrix} 0 & 1 \\ -1 & 0 \end{bmatrix}. \qquad (38)$$

Illustration

As an illustration, quarterly landings in Tamil Nadu, India during the period 1975-96 in respect of two fish species, *viz.* silverbellies' and croakers', is considered. These two species exhibit prominent seasonal fluctuations. The salient features of the data sets are having presence of trend and seasonality; hence BSMDS and BSMTS are the natural choice for modelling of these data sets. These models are fitted to the data sets using STAMP. It is found that BSMDS performs better than BSMTS for describing silverbellies data while it is the other way for describing croakers' data. Accordingly, parameter estimates and goodness of fit statistics for the best models from STM family are reported in Table 10.4. It is also observed from the table that in respect of both the fish species, σ_w^2 is less than σ_ε^2, which shows that recent observations are given more weight in constructing a seasonal pattern for the forecast function. Estimates of slopes (β_t), for both data sets, being very close to zero, imply an almost horizontal trend. Further, estimates of σ_ξ^2 and σ_w^2 as zero show that, for the data sets under consideration, there is hardly any variability in the slope portion of trend and seasonal component over the years. Thus, the fluctuations in the data sets are primarily due to level portion of trend and irregular components. Subsequently, the analogue of BSMDS and BSMTS from ARIMA family, *viz.* MA (5) is fitted to the data sets using PROC ARIMA module available in SAS software. The AIC values for silverbellies' and croakers' data are found to be respectively 460.89 and 290.21. These clearly bring about the superiority of structural time-series models over their ARIMA counterparts. Finally, forecast values in respect of the two fish species for the next four quarters are reported in Table 10.4.

Table 10.4: Fitting of Structural Time-series Models

Parameter/Species	*Silverbellies*	*Croakers*
(i) Estimation of parameters for BSMDS/BSMTS		
μ_t	12.62	3.48
β_t	0.08	0.01
γ_1	0.30	0.69
γ_2	0.39	-0.30
γ_3	0.40	-0.41
γ_4	-1.09	0.02
σ_ε^2	8.59	1.04
σ_η^2	0.51	0.11
σ_ξ^2	0.00	0.00
σ_w^2	0.00	0.00
(ii) Goodness of fit statistics		
AIC	218.40	59.08

Contd...

Table 10.4–Contd...

Parameter/Species	Silverbellies	Croakers
(iii) Forecast values (in thousand tonnes)		
1997 (1st quarter)	13.00	4.18
1997 (2nd quarter)	13.18	3.21
1997 (3rd quarter)	13.27	3.10
1997 (4th quarter)	11.86	3.55

Discussion

It is hoped that, in due course of time, other research workers would also start applying this promising technique in favour of conventional ARIMA approach for describing and forecasting time-series data with marked seasonal fluctuations. However, there is need to study its multi-species extensions, as in a real–life situation, there are bound to be prey-predator, competition or symbiotic types of interactions for seasonally fluctuating fish species. Appropriate estimation procedures along with relevant software package need to be developed in order to apply "Multivariate structural time-series models" to real multi-species fish data. Such studies would go a long way in helping policy makers in framing appropriate fisheries management policies.

STM Model in the Presence of Auto-Correlated Error

Time series data collected on various fisheries data is quite likely having the presence of autocorrelations among successive observations. To model data of this kind, Harvey (1996) proposed Structural Time-series Modelling with autoregressive process (Harvey and Phillips, 1979). The essence of this approach is that it attempts to represent the main features of the given data. Components such as trends and cycles are or are not included in the model depending upon their presence or absence in the data; this ensures to produce a plausible model at the outset. Thus, in the presence of only trend (μ_t) and error component (ε_t), the "Structural time-series model" can be written as

$$Y_t = \mu_t + \varepsilon_t, \qquad t = 1, 2, \ldots\ldots, T \tag{39}$$

where, $\varepsilon_t \sim N(0, \sigma_\varepsilon^2)$. Further, first order autoregressive process is introduced in the model by putting

$$\mu_t = \varphi_1 \mu_{t-1} + w_t \tag{40}$$

where, φ_1 is the AR (1) parameter and $w_t \sim N(0, \sigma_w^2)$. Eq. (39) is called the "observation equation". Eq. (40) is called the "state equation", since it describes the manner in which the state of a system at time t, say μ_t, is determined from the state of the system, μ_{t-1}, at time t-1. In eqs. (39) and (50), ε_t and w_t can be thought of respectively as "observation noise" and "model noise"; the relative importance of these two noise components can be controlled by specifying values for the variances σ_ε^2 and σ_W^2. The initial value μ_0 is assumed to be a random variable with mean m_0 and variance σ_0^2. For the estimation of parameters in eqs. (39) and (40), these have to be put in state space form (Meinhold and Singpurwalla, 1983). In that situation, estimation of parameters in the state vector is carried out using Kalman filter, prediction and smoothing (Harvey, 1996). The hyper parameters σ_ε^2 and σ_W^2 are assumed to be known, otherwise these can be estimated using Prediction error decomposition (PED) method or by Expectation maximization (EM) algorithm (Shumway and Stoffer, 2000) The reduced form from ARIMA family corresponding to eqs. (39) and (40) is ARIMA (1, 0, 0) or AR (1) model.

Illustration

In India, export of marine products presently contributes to nearly 25 per cent of gross national product from agriculture. Of this, shrimp contribute a major share to the basket of foreign exchange earnings. Therefore, its efficient modelling and forecasting will go a long way in helping the policy makers to frame appropriate export policies. It is hoped that, with global liberalization and setting up of World Trade Organization, export of various fish and fish products would witness a phenomenal growth rate in future. Under this scenario, other fish species would also exhibit quantum jump in their exports and would become important from economic point of view. In view of all this, modelling and forecasting of export data for fifteen years' (1985-86 to 2004-2005) of all–India shrimp export (Table 10.5) is utilized for the present study. A perusal of data indicates that there are no cyclical fluctuations present in the data. Further, as the data is on an annual basis, seasonal component is also not present.

Table 10.5: All-India Export of Shrimp

Year	*Export (in thousand tonnes)*	*Year*	*Export (in thousand tonnes)*
1985–86	50.3	1986–87	49.2
1987–88	55.7	1988–89	56.8
1989–90	57.8	1990–91	62.4
1991–92	76.2	1992–93	74.4
1993–94	86.5	1994–95	101.8
1995–96	95.7	1996–97	105.4
.1997–98	100.7	1998–99	102.5
1999–00	110.3	2000–01	117.5
2001–02	121.1	2002–03	140.8
2003–04	123.9	2004–05	137.8

In the first instance, a structural time-series model given by eqs. (39) and (40) is fitted to the data using STAMP software package. Parameter estimates and goodness of fit statistics are presented in Table 10.6. Subsequently ARIMA (1, 0, 0), *i.e.* AR (1) model is fitted to the data set using Statistical Analysis System (1990) software package, Version 6.12. The estimate of AR (1) parameter is obtained as 1.0 while the goodness of fit statistics are:

$$AIC = 127.8, SBC = 135.6, SE = 10.288$$

Therefore, a reduction of 71 per cent in AIC is achieved by employing structural time-series model with AR (1) process. Thus, this model performs much better than its ARIMA counterpart for the data set under consideration. Finally, forecast values for the next five years based on structural time-series modelling is presented in Table 10.6.

Table 10.6: Summary Statistics for Structural Time-series Model

Parameter	*Estimates*
(i) Fitting of model	
Level (μ_t)	137.70
φ_1	0.78
σ_ε^2	28.85
σ_w^2	10.78
(ii) Goodnes of fit statistics	
AIC	74.8
SBC	82.4
SE	6.1
(iii) Forecast (in thousand tonnes)	
2007-08	152.4
2008-09	157.3
2009-10	162.1
2010-11	167.0
2011-12	171.9

Bayesian Non-linear State Space Modelling for Schaefer Model

State space models are powerful tools for modelling and forecasting dynamic systems (Meinhold and Singpurwalla, 1983). These models, in conjunction with Kalman filter, have been used in several disciplines (Harvey, 1996; Jones and Nicholls, 1991). However, Kalman filter is optimal only when the dynamic system is normal. State space models consist of two equations, *viz.* observation equation and state transition equation. It is generally assumed that these two equations are linear and the error terms follow normal distribution. However, in reality, functional forms are generally nonlinear and errors follow non-normal distributions (Sullivan,1992). Bayesian approach can easily handle realistic distributional assumptions as well as nonlinearities in state and observation equations (Carlin *et al.*, 1992).

In any fishery, it is of prime importance to have an idea of sustainable yield, which can be harvested in perpetuity. To this end, Schaefer model is extensively employed. This model provided an argument, which ultimately led the International Whaling Commission to ban harvesting of blue whales in the oceans.

Following Polacheck *et al.* (1993), Schaefer model can be written as

$$B_t = B_{t-1} + r\,B_{t-1}\,(1 - B_{t-1}/k) - C_{t-1}, \tag{41}$$

where, B_t, C_t, r and k are respectively biomass at the start of year t, catch during year t, intrinsic growth rate, and carrying capacity. It is assumed that the relative abundance index, I_t, is directly proportional to biomass, *i.e.*

$$I_t = q\,B_t, \tag{42}$$

where, q is the catchability parameter. Maximum sustainable yield (MSY), is rk/4 which occurs at $B_{MSY} = k/2$. The optimal fishing effort to reap this MSY occurs at $E_{MSY} = r/(2q)$. The uncertainty present in eqs. (41) and (42) can be handled by state space models. These relate time-series observations $\{ I_t \}$ to unobserved states $\{ B_t \}$ through a stochastic transition model for I_t given B_t. Here the states are assumed to follow a stochastic transition model. Following Meyer and Millar (1999), the state equations are :

$$P_1 \mid s^2 = \exp(u_1)$$

$$P_t \mid P_{t-1}, k, r, s^2 = \{P_{t-1} + r\,P_{t-1}\,(1 - P_{t-1}) - C_{t-1} \mid k\}\exp(u_t),\ t = 2, 3, \ldots, N, \tag{43}$$

and the observation equations are

$$I_t \mid P_t, q, \tau^2 = q\,k\,P_t \exp(v_t),\ t = 1, 2, \ldots, N. \tag{44}$$

Here $u_t \sim N(0, s^2)$ and $v_t \sim N(0, t^2)$, x | y is the conditional distribution of x given y. Joint prior density obtained by successively applying Bayes theorem is

$$P(k, r, q, s^2, \tau^2, P_1, P_2, \ldots, P_N) =$$

$$P(k)\,P(r)\,P(q)\,P(s^2)\,P(\tau^2)\,P(P_1 \mid s^2) \prod_{i=2}^{N} P(P_t \mid P_{t-1}, k, r, s^2). \tag{45}$$

The sampling distribution is

$$P(I_1, I_2, \ldots, I_N \mid k, r, q, s^2, \tau^2, P_1, P_2, \ldots, P_N) = \prod_{t=1}^{N} P(I_t \mid P_t, q, t^2). \tag{46}$$

Hence the joint posterior distribution is

$$P(k, r, q, s^2, \tau^2, P_1, P_2, \ldots, P_N, I_1, I_2, \ldots, I_N)$$

$$= P(k)\, P(r)\, P(q)\, P(s^2)\, P(\tau^2)\, P(P_1 \mid s^2) \prod_{i=2}^{N} P(P_t \mid P_{t-1}, k, r, s^2) \prod_{t=1}^{N} P(I_t \mid P_t, q, \tau^2). \quad (47)$$

The posterior distribution can be obtained by integrating eq. (47) with respect to the parameters. However, this is a very cumbersome task. Recently, a promising technique of Markov Chain Monte Carlo (MCMC) using Gibbs sampling has been developed which enables evaluation of multiple integrals (Gibbs *et al.*, 1996). After obtaining the posterior distribution, various statistics of interest, like posterior mean, posterior standard deviation can easily be computed.

As an illustration, data on annual shrimp landings and their corresponding fishing efforts, as reported by Showers (1999), is considered for the present study. The data is reproduced in Table 10.7 for ready reference. Bayesian inference using Gibbs sampling (BUGS) software package, a recently developed software package (Spiegelhalter *et al.*, 1996) by the Medical Research Council, Biostatistics Unit, Institute of Public Health, Cambridge, England is utilized for data analysis. BUGS is available free of charge from the website http://www.mrc-psu.cam.ac.uk/bugs/Welcomes.htm for the operating systems LINUX, DOS, Windows, Unix. WinBUGS, Version 1.2 for the windows platform is used here. However, for fitting BNS model to data through WinBUGS, a computer program is required. The BNS model is fitted to data by running 1,50,000 iterations leaving the initial 25,000 iterations ("burn-in" period). Of the remaining 1,25,000 iterations, BUGS package monitors parameter estimates, stores

Table 10.7: Annual Shrimp Landings and Corresponding Fish Effort

Year	*Catch (in tonnes)*	*Effort (Fishing Days)*	*Index (CPUE)*
1981	642	2293	280
1982	880	2424	363
1983	1196	3680	325
1984	2304	8170	282
1985	2448	8298	295
1986	1944	7254	268
1987	2630	14064	187
1988	2750	17628	156
1989	2870	18397	156
1990	3000	20833	144
1991	1241	8862	140
1992	2484	14494	171
1993	2425	12422	195
1994	2167	10547	206
1995	2420	11435	212
1996	2788	16426	170

every 100th value, and calculates summary statistics of the sampled values. For each parameter, like carrying capacity (k), intrinsic growth rate (r), MSY, EMSY, and state transition probability (P[.]), summary statistics, *viz.* mean, standard deviation (SD), Monte Carlo error (MC), 2.5 per cent Lower percentile (LP), median, 97.5 per cent Upper percentile (UP), iteration number (IN) and number of iterations (NOI) were computed and a sample printout is presented in Table 10.8. A perusal of this table shows that MSY and EMSY are respectively 3668 tonnes and 17390 fishing days.

Table 10.8: Summary Statistics for Fitting BNS Model to Shrimp Data

Parameter	*Mean*	*SD*	*MC*	*LP*	*Median*	*UP*	*IN*	*NOI*
k (in tonnes)	4218	920.1	36.54	256.8	4195	6068	25000	5000
r	0.3458	0.2328	0.0043	0.0510	0.2841	0.8941	30000	5000
MSY (in tonnes)	3668	2641	157.6	918.7	2934	10846	125000	5000
EMSY (fishing days)	17390	12032	322	2680	14790	45900	145000	5000

This methodology can be extended for fitting other fisheries surplus production models, like Fox and Pella-Tomlinson models. However, this is a very challenging task as the theory and the corresponding software packages needs to be developed. The possibility of using BUGS package in other disciplines also may be explored.

Time-varying State Space Regression Models in Fisheries

In fisheries, data on catch, effort and other parameters of interest are usually collected over a long period of time. Linear regression model is generally adopted for analysis of such time series data by assuming that the regression relationship is constant over time. This may not be true as structure of the regression equation changes over time. To deal with this situation, models with time-varying coefficients should be considered. To this end, time-varying state space modelling approach using Kalman filtering technique may be effectively used. Kalman filter is a recursive procedure to update the parameters of a regression model when new data are observed (Young, 1984). This approach gives an insight into how parameter estimates are revised in a Bayesian fashion when new information is available in a world of uncertainty (Kim and Nelson, 1999). The purpose of present section is to highlight importance of time-varying state space modelling approach with Kalman filtering by considering data on all-India marine fish production and marine products export.

Model Description

Consider a regression model, in which regression coefficients are time- varying with specific dynamics:

$$Y_t = X_t \beta_t + e_t, \tag{48}$$

$$\beta_t = F \beta_{t-1} + v_t, \qquad t=1,2,\ldots,T. \tag{49}$$

In the above, it is assumed that $e_t \sim N(0, R)$, $v_t \sim N(0, Q)$, $E(e_t v_s') = 0$. Further, Y_t is an n x 1 vector of variables observed at time t; β_t is a k x 1 vector of unobserved state variables; X_t is an n x k matrix of independent variables; F is a k x k unit matrix. If F is a diagonal matrix with all diagonal elements less than 1, each regression coefficient follows a stationary AR (1) process. Eq. (48) is called "measurement equation" since it describes the relation between observed variables (data) and unobserved state variables. Eq. (49) is called "transition equation" since it describes dynamics of state variables. Eqs. (48) and (49) together is called "State space model". This is a useful tool for expressing dynamical systems that involve unobserved state variables. Generalized least squares (GLS) estimation of the parameter is

$$\beta_{t|t} = (X_t' \Omega_t^{-1} X_t)^{-1} X_t' \Omega_t^{-1} Y_t \,, \tag{50}$$

where, $\beta_{t|t}$ refers to an estimate of β_t conditional on information up to time t. A major difficulty is that its application requires an inverse of a t x t matrix, which is extremely cumbersome specially when t is quite large. Kalman filter approach for state space models can be easily implemented without having to invert large matrices (Harvey, 1996).

Kalman Filter and Estimation of State Vector

Kalman filter is a recursive procedure for computing optimal estimate of unobserved state vector β_t, t = 1,2, T, based on the appropriate set, assuming that F, R and Q are known. It provides minimum mean squared error estimate of β_t. Basic filter refers to an estimate of β_t based on information available up to time t, and smoothing refers to an estimate of β_t based on all available information in the sample through time T. Assuming that X_t is available at the beginning of time t and a new observation of Y_t is made at the end of time t, basic Kaman filter consists of following two steps:

Prediction

At the beginning of time t, an optimal predictor of Y_t based on all available information up to time t-1 is $Y_{t|t-1}$. Estimate (expectation) of β_t conditional on information up to time t-1 is

$$\beta_{t|t-1} = E[\beta_t \mid \psi_{t-1}] = F\,\beta_{t-1|t-1} \,, \tag{51}$$

where, ψ is the information set. Covariance matrix of β_t conditional on information up to time t -1 is

$$P_{t|t-1} = F\, P_{t-1|t-1}\, F' + Q \;. \tag{52}$$

Prediction error is computed as

$$\eta_{t|t-1} \equiv Y_t - Y_{t|t-1} = Y_t - X_t\, \beta_{t|t-1}, \tag{53}$$

and its conditional variance is

$$f_{t|t-1} \equiv E\,(\eta^2_{t|t-1}) = X_t\, P_{t|t-1}\, X'_t + R \;. \tag{54}$$

Updating

Once Y_t is realised at the end of time t, prediction error is calculated. It contains new information about β_t beyond that contained in $\beta_{t|t-1}$. Thus

$$\beta_{t|t} = \beta_{t|t-1} + K_t\, \eta_{t|t-1}, \tag{55}$$

and its covariance is

$$P_{t|t} = P_{t|t-1} - K_t\, X_t\, P_{t|t-1}, \tag{56}$$

where, $K_t = P_{t|t-1}\, X'_t\, f'_{t|t-1}$ is the Kalman gain, which determines the weight assigned to new information about β_t contained in prediction error. Given initial values, $\beta_{o|o}$ and $P_{o|o}$, six equations, *viz.* eqs. (51) to (56) can be iterated for t = 1, 2, ..., T. This provides a minimum mean square estimate of β_t, t = 1, 2, ..., T given information up to time t-1 or t. Smoothing $\beta_{t|T}$ yields a more accurate inference on β_t, since it uses more information than the basic filter. Hence, iterating backward for t = T-1, T-2, 1, estimate of β_t conditional on information up to T (whole sample) is obtained as

$$\beta_{t|T} = \beta_{t|t} + P_{t|t}\, F'\, P^{-1}_{t+1|t}\, (\beta_{t+1|T} - F\beta_{t|t}), \tag{57}$$

and its covariance is

$$P_{t|T} = P_{t|t} + P_{t|t}\, F'\, P^{-1}_{t+1|t}\, (P_{t+1|T} - P_{t+1|t})\, P^{-1}_{t+1|t}\, F'\, P'_{t|t}. \tag{58}$$

Illustration

Data on all-India marine products export and marine fish production during the period 1970-71 to 2004-05 is considered. The data is reproduced in Table 10.10 for ready reference. Evidently, marine products export during a particular year depends on marine fish production of that year. Therefore, marine fish production (X) may be treated as an 'independent variable' while marine products export (Y) as 'dependent variable'. It may be mentioned that, in subsequent analyses, models are fitted to data after ignoring last observation. The value corresponding to this observation is then

Table 10.9: All-India Data on Marine Fish Production and Marine Products Export (in thousand tonnes)

Year	*Marine Fish Production (X)*	*Marine Products Export (Y)*
1970–71	1086	35.9
1971–72	1137	35.5
1972–73	1033	38.9
1973–74	1241	52.3
1974–75	1244	45.1
1975–76	1434	54.5
1976–77	1410	66.8
1977–78	1310	66.0
1978–79	1430	86.9
1979–80	1492	86.4
1980–81	1555	75.6
1981–82	1445	70.1
1982–83	1427	78.2
1983–84	1519	92.7
1984–85	1698	86.2
1985–86	1716	83.7
1986–87	1713	85.8
1987–88	1658	97.2
1988–89	1817	99.8
1989–90	2275	110.2
1990–91	2300	139.4
1991–92	2447	171.8
1992–93	2576	208.6
1993–94	2649	244.0
1994–95	2692	307.3
1995–96	2707	296.3
1996–97	2967	378.2
1997–98	2950	385.8
1998–99	2696	388.0
1999–00	2852	398.7
2000–01	2811	421.1
2001–02	2830	424.3
2002–03	2990	498.0
2003–04	2941	409.7
2004–05	2778	450.1

predicted using fitted models and an assessment of goodness of fit of models is made on the basis of discrepancy. In the first instance, a linear regression model with additive errors, assumed to be independently distributed, is fitted to the data using SAS software. Run test for testing independence of errors indicates that the calculated value of Z (-4.275) lies in the rejection region as the tabulated value at 5 per cent level is -1.645, and so errors are not independent. Further, Durbin–Watson (DW) statistic value of 0.138 indicates presence of autocorrelation of order one among successive observations. Therefore, error term in eq. (51) is assumed to be autoregressive (AR) of order one for subsequent analysis and this model is referred to as Model I Estimation of AR (1) parameter for this model is carried out using Melard's algorithm (Melard, 1984) and the results are presented in second column of Table 10.2. As the data pertains to a large period of time, assumption of constancy of regression parameter appears doubtful. Therefore, time–varying state space regression model, referred to as Model II, is fitted to the data using Structural Time Series Analyser, Modeller and Predictor (STAMP) software package. It may be pointed out that this type of analysis can also be carried out using SsfPack2.2 (Koopman *et al.*, 1999) software package. The results are presented in third column of Table 10.10. A perusal of goodness of fit statistics for the two models indicates that Model II provides a substantial improvement over Model I. Also forecast of marine products export for 2007-08 obtained using models I and II are respectively 478.97 and 465.85 thousand tonnes against actual value of 450.1 thousand tonnes, reflecting thereby superiority of Model II over Model I. Time–varying state space model using Kalman filtering technique should be adopted for modelling fisheries data.

Table 10.10: Fitting of Models I and II to the Data

Parameter/Statistics	*Model I*	*Model II*
$\hat{\beta}$ (S.E.($\hat{\beta}$))	0.092(0.035)	0.074(0.048)
AR(1)	0.615	—
DW	1.340	2.173
$R \equiv \sigma_e^2$	-	52.623
$Q \equiv \sigma_v^2$	-	48.917
Goodness of fit statistics		
Log Likelihood	–156.76	-82.97
Mean Square Error	32.54	18.98
AIC	287.73	172.85
SBC	295.62	180.47

Further Scope of Research

Techniques such as Generalized Kalman Filter (GKF) and Extended Kalman Filter (EKF) involving nonlinear state space models may be utilised in research problems. Software for fitting STM models for non-linear state space models is not readily available and hence has to be developed. More emphasis may be made on "Structural time series modelling" approach in favour of ARIMA time–series modelling approach for development of models using time-series data. Modelling and forecasting of data dealing with single-species as well as multi–species is highly desirable. Researchers should apply STM models with multivariate extension in some real-life situations, *e.g.* in fisheries, entomology, there are interacting species exhibiting cyclical fluctuations and having prey-predator, competition, or symbiotic types of interaction for seasonally fluctuating fish species. Appropriate estimation procedures along with relevant computer programs to handle such situations have to be developed in order to apply "Multivariate structural time series models" to real-life data. STM modelling can be utilised for describing seasonal and cyclical fluctuations in India's rainfall data as the rainfall fluctuates due to erratic monsoon influenced by global climate change. Such studies would go a long way in helping policy makers in framing appropriate modelling and forecasting policies.

Conclusion

STM modelling and forecasting methodology can be effectively utilized in agriculture where lots of data on various agricultural commodities are collected over time. In fisheries also, data on marine products export, data on catch, effort and other parameters of interest etc. are usually over a long period of time and at regular intervals. Modelling these time series data can be effectively carried out using STM procedure and forecasting can also be carried out with better accuracy. STM methodology can be effectively utilized in fisheries and agriculture where observations are collected over regular intervals and where linear regression modelling procedure is utilized for forecasting these parameters. Since, advanced statistical software packages are available, STM modelling methodology can be efficiently utilized, when there are more number of data points and when there are more number of explanatory variables. In Indian agriculture and fisheries also, STM modelling modelling methodology can be effectively utilized since accurate modelling and forecasting would help policy planners in framing suitable policies.

References

Box, G.E.P., Jenkins, G. M. and Reinsel, G. C. (1994). *Time series analysis : Forecasting and control,* 3rd edition. Prentice Hall, New Jersey.

Carlin, B.P. Polson, N.G. and Stoffer, D.S. (1992). A Monte Carlo approach to no normal and nonlinear state-space modelling. J. Am. Stat. Assoc. 87 : 493-500

de Jong P. (1988). The likelihood for a state space model. *Biometrika* 75:165-169.

Gilks, W.R., Richardson, S. and Spiegelhalter, D.J. (1996). Markov-chain Monte Carlo in practice. Chapman and Hall, London, U.K.

Harvey A.C. (1996). *Forecasting, Structural Time Series Models and the Kalman Filter.* Cambridge Univ. Press, U.K.

Harvey A C and Durbin J. (1986). The effects of seat belt legislation on British road casualties : a case study in structural time series modelling. *Journal of Royal Statistical Society,* A 149 : 187-227.

Koopman S J, Harvey A C, Doornik J A and Shephard N. (2000). *STAMP : Structural Time Series Analyser, Modeller and Predictor*. Timberlake Consultants Press.

Meinhold, R. J. and Singpurwalla, N.D. (1983). Understanding Kalman filter. *Amer. Statistician*, 37, 123-127.

Meyer, R. and Millar, R.B. (1999). BUGS in Bayesian stock assessments. Can. J. Fish. Aquat. Sci. 56, 1078-1087.

Polacheck, T., Hilborn, R. and Punt, A.E. (1993). Fitting surplus production models : Comparing methods and measuring uncertainty. Can. J. Fish. Aquat. Sci. 50, 2597-2607.

Ravichandran, S. and Prajneshu (2001). "State space modelling versus ARIMA Time Series Modelling" in Journal of Indian Soc. of Ag.Statistics, 54(1), 2001, pp. 43-51.

Ravichandran, S. and Prajneshu (2002). "Dynamical modelling and forecasting of India's foodgrain production", Proc. Nat. Acad. Sci. India, Vol. 72, B(I).

Ravichandran, S. and P.Muthuraman (2006). "Structural time-series modelling and forecasting India's rice production", Agricultural situation in India, pp.773–776.

Ravichandran, S., Prajneshu and Savita Wadhwa (2002). "Structural Time-Series Models for Describing Cyclical Fluctuations", J.of Ind. Soc. of Agrl.Stat., 55(1), pp. 70-78.

Ravichandran, S. and P. Muthuraman (2003). "Structural time series models for describing sunflower yield", Indian Journal of Agricultural Sciences.

Ravichandran, S. and Prajneshu (2004). "Structural time-series models for describing seasonal fluctuations in fisheries", Indian J. of Animal Sciences.

Selukar, R. (2003). *PROC UCM: SAS Institute, Inc.*, pp. 1497-1563

Showers, P.A.T. (1999). Escalation in shrimp production in the Sierra Leone industrial fishery, Naga, 22, 29-31.

Sullivan, P.J. (1992). A Kalman filter approach to catch-at-length analysis. Biometrics, 48, 237-257.

Shumway, R.H. and Stoffer, D.S. (2000). *Time series analysis and its applications.* Springer Verlag, New York.

Spiegelhalter, D.J., Thomas, A., Best, N. and Gilks, W.R. (1999). WinBUGS 1.2 : Bayesian inference using Gibbs sampling, User manual. Medical Research Council Biostatistics Unit, Institute of Public Health, Cambridge, England.

Modelling, Forecasting, Artificial Neural Network and Expert System in Fisheries *Pages 148–154*
Editors: **Ajit Kumar Roy & Niranjan Sarangi**
Published by: **DAYA PUBLISHING HOUSE (2009)**

Chapter 11

Tropho-Dynamic Modelling for Ecosystem Based Fisheries Management

Preetha Panikkar and M. Feroz Khan

An ecosystem is defined as "a spatially explicit unit of earth that includes all of the organisms, along with all the components of the abiotic environment within its boundaries" (Likens, 1992). All these components are connected in a complex food web by evolving interactions. A comprehensive understanding of the structure, function, and regulation of major ecosystems is necessary to face the world's ever-growing environmental problems (Gaedke 1995). It also becomes important to understand what impacts an ecosystem can tolerate before major structural changes occur, and how reversible these changes are. In this respect, improved understanding of ecosystem dynamics is critical to predict and manage the consequences of environmental variability and human impacts, such as those induced by fisheries, an activity targeting specific species and size-classes. Ecosystem indicators can be used to monitor aggregate changes in fish populations and communities in response to natural and human-induced perturbations. Such indicators could be used as effective tool for fisheries management. Ecosystem-based fishery management is a new direction for fishery management, essentially reversing the order of management priorities to start with the ecosystem rather than the target species. The overall objective of ecosystem based fishery management is to sustain healthy aquatic ecosystems and the fisheries they support.

The emerging shift of fisheries research from single-species analysis towards an ecosystem-based approach requires tools that explicitly account for ecological interactions, especially those of a trophic nature. Mass-balance biomass model is such a tool (Ecopath approach) used globally as an efficient and useful method to systematically describe ecosystems and to explore their properties (Pauly and Christensen 1996; Savenkoff *et al.*, 2001). They constitute a simple approach to represent the complexity of an ecosystem, and they involve a mass-balance description of trophic interactions between all the functional groups of the ecosystem. Such scientific tools will provide valuable information on the health of the ecosystem as well as the capacity to support biological production and trophic functioning.

Ecopath with Ecosim

The Ecopath with Ecosim (EwE) modelling approach combines software for ecosystem trophic mass balance (biomass and flow) analysis (Ecopath), with a dynamic modelling capability (Ecosim) for exploring past and future impacts of fishing and environmental disturbances. It has a very elaborate user interface that eases a variety of data management chores and calculations that are a cumbersome but necessary part of any endeavor to systematically examine an ecosystem. EwE is coming to be widely used as a tool for analysis of exploited aquatic ecosystems, having reached 2500 registered users in 120 countries.

The Ecopath approach and software have been under continuous development since the late 1980s (Christensen and Pauly 1992), with Ecosim emerging in 1995 (Walters *et al.*, 1997, 2000). Depending on the system under consideration and the information available, the Ecopath with Ecosim applications can be envisaged in relation to fisheries management. In many cases, the period considered will be a typical season, or a typical year, but the state and rate estimates used for model construction may pertain to different years. Models may represent a decade or more, during which little changes have occurred. When ecosystems have undergone massive changes, two or more models may be needed, representing the ecosystem before, during, and after the changes.

The foundation of the EwE suite is an Ecopath model (Christensen and Pauly 1992, Pauly *et al.*, 2000), which creates a static mass-balanced snapshot of the resources in an ecosystem and their interactions, represented by trophically linked biomass pools. The biomass pools consist of a single species, or species groups representing ecological guilds. Pools may be further split into ontogenic (juvenile/adult) groups that can then be linked together in Ecosim. Ecopath data requirements are relatively simple, and generally already available from stock assessment, ecological studies, or the literature: biomass estimates, total mortality estimates, consumption estimates, diet compositions, and fishery catches.

The Ecopath Model

Ecopath parameterizes models based on two master equations, one to describe the production term and one for the energy balance for each group. The first Ecopath equation describes how the production term for each group (i) can be split in components. This is implemented with the equation,

Production = catches + predation mortality + biomass accumulation + net migration + other mortality;

i.e., $Pi = Yi + Bi.M2i + E\,i + BA\,i + M0\,i$

where, Pi is the total production rate of (i), Yi is the total fishery catch rate of (i), M2i is the total predation rate for group (i), E i the net migration rate (emigration – immigration), BA i is the biomass accumulation rate for (i), while M0i is the 'other mortality' rate for (i). Of the terms in the equation above, the production rate, Pi, is calculated as the product of Bi, the biomass of (i) and Pi/Bi, the production/biomass ratio for group (i). The Pi/Bi rate under most conditions corresponds to the total mortality rate, Z, (Allen, 1971), commonly estimated as part of fishery stock assessments. The 'other mortality' is a catch all term including all mortality not elsewhere included, *e.g.*, mortality due to diseases or old age.

A box (group) in an Ecopath model may be a group of (ecologically) related species, a single species, or a single size/age group of a given species. In a model, the energy input and output of all living groups must be balanced. The first Ecopath equation includes only the production of a box.

When balancing the energy balance of a box, other flows should be considered. After the missing parameters have been estimated so as to ensure mass balance between groups energy balance is ensured within each group using the second equation

Consumption = production + respiration + unassimilated food

$$Qi = Pi + Ri + Ui$$

where, Qi = consumption of group i, Ri = respiration of group i and Ui = unassimilated food of group i. If the total consumption of each group did not balance the production and loss terms (respiration and unassimilated biomass), then the biomass of group i would change over time. This equation is in line with Winberg (1956) who defined consumption as the sum of somatic and gonadal growth, metabolic costs and waste products.

For parameterization Ecopath sets up a system with (at least in principle) as many linear equations as there are groups in a system, and it solves the set for one of the following parameters for each group:

- ✰ Biomass;
- ✰ Production/biomass ratio;
- ✰ Consumption/biomass ratio; or
- ✰ Ecotrophic efficiency

while the other three parameters along with following parameters must be entered for all groups:

- ✰ Catch rate;
- ✰ Net migration rate;
- ✰ Biomass accumulation rate;
- ✰ Assimilation rate; and
- ✰ Diet compositions

The main aim of tropho-dynamic modelling is to regulate fishing mortality rates over time so as to achieve economic, social, legal and ecological sustainability objectives. Dynamic modelling and assessment provides insight about how high these mortality rates should be, and how they should be varied over time from past over-fishing. We cannot expect models to provide very precise estimates of optimum fishing mortality rates, but we should at least be able to define reasonable and prudent ranges for the rates. The impacts of alternative time patterns of fishing mortalities can be explored using two different approaches in Ecosim. Dynamic simulations can also simulate a regime shift by changing primary productivity and diet compositions. Assumptions about the available flow of food from one group to another can be evoked in both dynamic and equilibrium models to simulate 'top-down' or ' bottom-up' control.

Network Analysis

Once a balanced Ecopath model is obtained, the flows of biomass among the groups and higher-order indices of ecosystem functioning can be interpreted. Ecopath provides the flows of biomass among groups that satisfies the steady-state condition, and that are also consistent with the inputted values of production (*P*), biomass (*B*), and consumption (*Q*). Several network analysis indices are also produced by Ecopath, which are useful for determining an ecosystem's structure, maturity, and stability (Odum, 1969; Ulanowicz, 1986). These indices are total system throughput (*T*), ascendency (*A*), system capacity (*C*), and system overhead, which are based on ascendency and capacity. Total system throughput (*T*) represents the entire amount of biomass flow within the system. Ascendency (*A*) measures the structure of an ecosystem in terms of the amount and organization of biomass flow within the system. Based upon Odum's (1969) interpretation of the attributes of ecosystems, more speciation, finer specialization, longer retention, and more cycling within the system indicate that an ecosystem is more mature. The upper limit to ascendency is the development capacity (*C*) of the ecosystem. System overhead is the difference between capacity and ascendency. System overhead is the upper limit to how much ascendency can increase to counteract unexpected perturbations. Higher overhead indicates that a system has a larger amount of energy reserves with which it can react to perturbations, so that the system should be more able to maintain stability when perturbed.

Ecopath gives the mixed trophic impacts within the ecosystem, which is shown in Mixed Trophic Impact (MTI) allows assessing the impact that changes in biomass of a group will have on the biomass of the other groups in an ecosystem trophically. Mixed trophic impact may help to explain short-term changes (Mavuti *et al.*,1996) but it cannot be taken as an instrument for making medium or long-term predictions.

Ecosystem 'Health'

The health status of a patient can often be captured with a single parameter, the temperature. Many have tried to find an index with similar ability to describe the health of an ecosystem to avoid the insurmountable task associated with bottom-up approach summing up the health of all ecosystem components, but a clear indicator has not appeared. The effort has led to development and description of a variety of system indicators, typically though with a given researcher exploring only one or a few of the potential indicators and on one or a few systems only.

One area of research of this approach relates to ecosystem maturity, a potential descriptor of ecosystem health (Christensen, 1995). Odum (1969, 1971) described how ecosystems develop over time in a non-deterministic way. Implications of this include that in a more mature system all niches should tend to be filled; that a larger part of the energy flows should be through detritus-based food webs; that primary production should be more efficiently utilized; that the total system biomass/energy throughput ratio should be higher; etc. An ecosystem attains maturity after several ecological successions and hence development and maturity of an ecosystem stand in opposition to each other. A mature ecosystem has the capacity to withstand perturbations caused by human beings or nature more than an immature ecosystem.

When ecosystems are disturbed, notably by fishing, we expect their maturity to decrease. This was indicated by the findings of Christensen (1995), who used a series of indicators to rank a large number of ecosystem representations after maturity, and concluded that the ranking obtained was in agreement with the expected state of maturity. Some of the indices for determining the maturity of an ecosystem are as follows.

1. Respiration/Assimilation ratio can be more than 1
2. Production/Respiration ratio is always less than 1
3. Primary production/Respiration ratio is greater than 1 in the early developmental stages of an ecosystem. In mature system, the value is around 1 but in polluted system the ratio is less than1.
4. Primary production/Biomass ratio is less than 1 in immature system since the biomass accumulates
5. Net system production is the difference between total primary production and total respiration. In immature system, the production is large, but in mature systems, it is close to zero.
6. Connectance index is the ratio of the actual links to the number of possible links in a given food web. Food chain structure changes from linear to web-like as systems mature.

Ecosim: Time-dynamic Simulation

The Ecosim part of EwE software facilitates dynamic model progression over time, given specified changes in parameters. Ecosim provides a very useful approach for evaluating the multi-species impacts of multi-species management approaches. Ecosim bases the crucial assumption for prediction of consumption rates on a simple Lotka-Volterra or 'mass action' assumption, modified to consider 'foraging arena' properties.

Ecosim incorporates the foraging arena assumption through a parameter called vulnerability (v), which describes how much the predation mortality for a given prey can increase if the predator abundance is increased, When the predator is at its carrying capacity with regard to the given prey, the predation mortality cannot be increased any further (v=1), and an increase in predator abundance, (*e.g.*, due to good recruitment) will be compensated for by a decrease in predator consumption rates. This in turn will result in lower predator production, and the predator abundance will move back toward its carrying capacity.

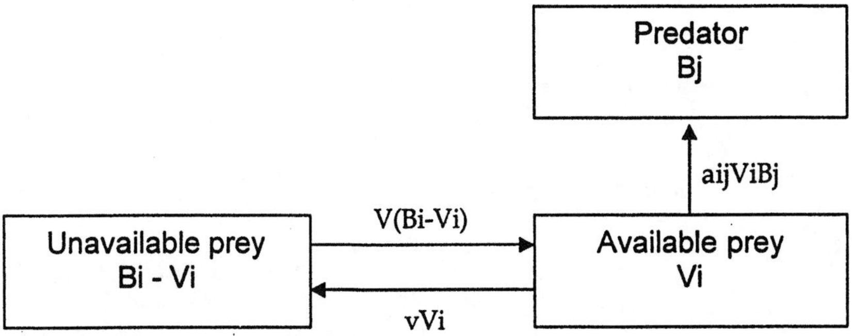

Simulation of flow between available (Vi) and unavailable (Bi–Vi) prey biomass in Ecosim. aij is the predator search rate for prey i, v is the exchange rate between the vulnerable and not-vulnerable state. Fast equilibrium between the two prey states implies $Vi = vBi/(2v + aBj)$. The fig. shown is based on Walters et al. (1997).

The basics of Ecosim consist of biomass dynamics expressed through a series of coupled differential equations. The equations are derived from the Ecopath master equation (Equation 1), and take the form

$$dBi/dt = gi \Sigma C ji - \Sigma Cij + I i - (Mi + Fi + ei) Bi$$

where, dBi/dt represents the growth rate during the time interval dt of group (i) in terms of its biomass, Bi, gi is the net growth efficiency, Mi the non-predation ('other') natural mortality rate, Fi is fishing mortality rate, ei is emigration rate, Ii is immigration rate. The consumption rates, Cji, are calculated based on the 'foraging arena' concept, where Bi's are divided into vulnerable and invulnerable components (Walters *et al.*, 1997), and it is the transfer rate (vij) between these two components that determines if control is top-down (*i.e.*, Lotka-Volterra), bottom-up (*i.e.*, donor-driven), or of an intermediate type. The set of differential equations is solved in Ecosim using an Adams-Basforth integration routine or (if selected) a Runge-Kutta 4 th order routine.

Conclusion

Mass-balance Ecopath and Ecosim modelling has advantages in making clear the impacts of harvest, comparing the effects of different gear types, and in being able to provide estimates of unknown biomasses. It can therefore validate anecdotal information on presence or absence, and relative abundance of fish species. The Ecopath with Ecosim software has not only increased the quantitative power of approach, but has allowed qualitatively new questions to be asked. The package is extremely useful and, if critically applied, can provide very useful information complementary to existing assessment/simulation methods. Its great value lies in the fact that it will enable at least some consideration to be given, within a structured framework, to ecosystem and spatial effects, in a wide range of cases where, until now, such features have been essentially overlooked.

References

Allen, R. R., 1971. Relation between production and biomass. *J. Fish. Res. Board Can.* 28:1573-1581.

Christensen, V. and Pauly, D., 1992. "ECOPATH II–a software for balancing steady-state ecosystem models and calculating network characteristics." *Ecological Modelling* 61:169-185.

Christensen, V., 1995. "Ecosystem maturity–towards quantification." *Ecological Modelling* 77:3-32.

Gaedke, U., 1995. A comparison of whole-community and ecosystem approaches (biomass size distributions, food web analysis, network analysis, simulation models) to study the structure, function and regulation of pelagic food webs. *Journal of Plankton Research* 17:1273-1305.

Likens, G. E., 1992. Beyond the shoreline: A watershed-ecosystem approach. *Verh. Internat. Verein. Limnol.* 22:1–22.

Mavuti, K. M., Moreau, J., Munyandorero, J. and Plisnier, P. D., 1996. Trophic relationships in two shallow lakes of Africa, Lake Naivasha (Kenya) and Lake Iherna (Rwanda), comparison with Lake George. *Hydrobiologia* 321: 89-100.

Odum, E. P., 1969. "The strategy of ecosystem development." *Science* 164:262-270.

Odum, E. P., 1971. Fundamentals of ecology. W.B.Saunders Co., Philadelphia. 574p. Life history strategies in aquatic ecosystems: Ecosim II. *Ecosystems* 3:70-83.

Pauly, D. and Christensen, V., 1996. Mass balance models of North eastern Pacific ecosystem. Fisheries Centre Research Reports,4 (1): 131p.

Pauly, D., Christensen, V. and Walters, C. J., 2000. Ecopath, Ecosim and Ecospace as tools for evaluating ecosystem impact of fisheries. *ICES J. Mar. Sci.* 57:697-706.

Savenkoff, C., Vézina, A. F. and Bundy, A., 2001. Inverse analysis of the structure and dynamics of the whole Newfoundland-Labradoe shelf ecosystem. *Canadian Technical Report of Fisheries and Aquatic Sciences* No. 2354.

Ulanowicz, R.E., 1986. Growth and development: ecosystem phenomenology. Springer Verlag, New York, 203 p.

Walters, C. M., Christensen, V. and Pauly, D., 1997. "Structuring dynamic models of exploited ecosystems from trophic mass-balance assessments.' *Reviews in Fish Biology and Fisheries*, 7:139-172.

Walters, C., Pauly, D., Christensen, V. and Kitchell, J. F., 2000. "Representing density dependent consequences of life history strategies in aquatic ecosystems: EcoSim II." *Ecosystems*, 3:70-83.

Winberg, G. G., 1956. Rate of metabolism and food requirements of fishes. *Transl. Fish. Res. Board Can.* 253.

Modelling, Forecasting, Artificial Neural Network and Expert System in Fisheries
Editors: Ajit Kumar Roy & Niranjan Sarangi
Published by: DAYA PUBLISHING HOUSE (2009)

Pages 155–178

Chapter 12

Forecasting of Fish Production of India by Culture Environment and Species

A.K. Roy and Nirupama Panda

Introduction

Fisheries resources in India constitute a coastline of 8,129 km, Exclusive Economic Zone of 2.02 million sq. km, Continental Shelf of 0.506 million sq. km, Rivers and Canals of 1,97,024 km, Reservoirs of 3.15 million ha, Ponds and Tanks of 2.35 million ha, Oxbow lakes and derelict waters of 1.3 million ha, Brackishwaters of 1.24 million ha and Estuaries of 0.29 million ha. At present, the total fish production is about 6.4 mmt. out of which; inland fish production is 3.4 mmt. and marine is 3.0 mmt. Although there is a potential of fish production of 8.4 mmt. There are 1,070 hatcheries producing fish seed production of 21,000 million fry. Besides 429 FFDA[5] and 39 BFDA[5] are operating in the country (Anon, 2006 a). Ponds and tanks are the prime resources for aquaculture, however only about 0.8–0.9 million ha is used for aquaculture currently. Indian aquaculture has demonstrated a six and half fold growth over the last two decades with freshwater aquaculture contributing over 95 per cent of the total aquaculture production. Three Indian major carps namely catla (*Catla catla*), rohu (*Labeo rohita*) and mrigal (*C. mrigala*) contribute the bulk of production over 1.8 million tons (FAO, 2005).

Fish is an important source of protein especially in developing countries. Fish accounts for 20 per cent of animal derived protein in low-income, food deficit countries compared with 13 per cent in the industrialized countries (L. Christopher *et al.*, 2002). Despite the rapid growth in fish production in India, the level of per capita consumption is still much lower than the actual requirement. Hence, the expansion of fish demand is likely to continue for some more years. The consumption of fish such as carps and other low-valued fish have shown a higher rate over the recent past. Accordingly, adequate supply of these low valued species is needed to cope with the requirement for the present and the future as well. Rapid population growth along with increase in income level and average per capita fish consumption leads to soaring demand for food fish production in India. Though production from capture fisheries has leveled off, production of fish from aquaculture has exploded in past 55 years as evidenced from the time series data of FishStat Plus, FAO. But, the question arises whether sufficient quantity from aquaculture would be produced to meet the anticipated requirement of people during the next 15 years or so? From the declining growth rate in capture fisheries for the period from 1950 to 2004 compared to that of aquaculture, it is presumed that the estimated production from capture fisheries in 2020 will be far less than aquaculture production. Over exploitation of capture fisheries and its declining growth rate suggest for projection of fish production from capture fisheries. The production of high value crustaceans always supports for export earning and thus adds to the revenue basket of the country. Accordingly, adequate production of crustaceans is required to increase the level of national revenue.

Survey of literature reveals that, some attempts have been made to forecast fish production and consumption (Delgado *et al.*, 2003; Wijkstrom, 2003 and Ye, 1999) Forecasting of species wise freshwater aquaculture production was done in China and the factors responsible for increase in production were identified (Cremer, *et al.* 1998). Extrapolation methods have been widely introduced for development of forecasting fisheries and the interval method of expression is used in the forecast indices (Vacek, J. 1973). Non-linear statistical models for pre-harvest forecasting of fish production from inland ponds in India *viz.* rohu, mrigal and common carp were fitted to forecast fish weight at the time of harvest after 12 months of stocking (Walia and Jain, 1998). Attempts were also made to apply the non-parametric regression with auto correlated errors methodology for describing India's marine fish production data for the period 1971 to 2000 (Chandran and Prajneshu, 2005). A report on estimate of the inland aquaculture production is also available (K. Gopakumar, 2003). Projection of fish demand by species groups was made for the year 2015 by using demand parameters based on household dietary pattern and fish consumption by species in the districts of Andhra Pradesh, Haryana, Karnataka, Uttar Pradesh, West Bengal and Orissa in the year 2002 (Kumar, *et al.*, 2005). But at international level, a few studies are available (De Silva, 2001; Delgado *et al.*, 2003; Josupeit *et al.*, 2004; Ye, 1999).

Although some relevant reports are available in this field, but so far, no scientific study has been undertaken to estimate group wise and species wise fish production in India. Therefore, it is attempted to project group wise and species wise fish production from both sectors of fisheries *viz.* capture and culture in 2020 using the time series data of FishStat, FAO (version 2.3). Preliminary studies on group wise and species wise estimates of total fish production and freshwater aquaculture production have been recently reported to forecast capture fish production in India in 2020 (Roy and Panda, 2007a, 2007b, 2007c). The time series data of FishStat Plus, FAO for the period from 1950 to 2004 includes total fish production, aquaculture production and capture production both in terms of quantity and value. The definitions and terminologies employed by the FAO (FAO, 2007) are as follows:

Aquaculture: The farming of aquatic organisms including fish, molluscs, crustaceans and aquatic plants is called Aquaculture. Farming implies some form of intervention in the rearing process to enhance production, such as regular stocking, feeding, protection from predators etc. Farming also implies individual or corporate ownership of the stock being cultivated. For statistical purposes, aquatic organisms harvested by an individual or corporate body that has owned them throughout their rearing period contribute to aquaculture, while aquatic organisms that are exploitable by the public as a common property resource, with or without appropriate licenses, are the harvest of fisheries.

- ☆ *Aquaculture Production*: Aquaculture production refers to output from aquaculture activities that are designated for final harvest for consumption or other purposes (*e.g.* ornamental purposes). Aquaculture production is also reported by three culture environments, namely fresh water, brackish water and marine water.
- ☆ *Freshwater* is water with a consistently negligible salinity.
- ☆ *Brackish water* is water that may reach high salinity levels, but this is not constant. It is usually characterized by regular daily and seasonal fluctuations in salinity due to freshwater and full strength marine water influxes. Enclosed coastal and inland water bodies in which the salinity is greater than fresh water but less than marine water are also regarded as brackish.
- ☆ *Marine water* is coastal and offshore water in which the salinity is maximal and not subject to significant daily or seasonal variations.

Total Fish Production

The time series data on fish production from 1950-2004 shows that total fish production has increased from 0.73 mmt in 1950 to 6.10 mmt in 2004 recording 8.35 times increase over the period 1950. Total Fish Production comprises Aquaculture production (2.47 mmt, 40.5 per cent) and Capture production (3.63 mmt, 59.5 per cent) in 2004. The projected figure for 2020 is estimated as 8.79 mmt which will be the sum total of aquaculture production of 5.84 mmt (66.4 per cent) and capture production of 2.95 mmt (33.6 per cent) in 2020, which compares favaourably with the projected production of 8.00 mmt in 2020 under most likely (baseline) estimate of International Food Policy Research Institute (IFPRI, 2002). The percent increase of total fish production in 2004 is 735.62 per cent over the production in 1950 whereas culture production has increased by 12250 per cent over the production of 0.02 mmt in 2004 to 2.47 mmt in 2004, the corresponding annual per cent increase for capture fishery is 222.73 per cent.

Table 12.1: Fish Production during 1950–2004 and Forecast 2020

	1950	1960	1970	1980	1990	2000	2001	2002	2003	2004	Forecast (2020)
Aquaculture production	0.02	0.04	0.12	0.37	1.02	1.94	2.12	2.19	2.31	2.47	5.84
Capture production	0.71	1.12	1.64	2.08	2.86	3.73	3.82	3.74	3.72	3.63	2.95
Total Fish Production	0.73	1.16	1.76	2.45	3.88	5.67	5.94	5.93	6.03	6.10	8.79

Source: Roy and Panda, 2007a.

From the figures in the Table 12.1, it is significant to note that the ratio of fish production from capture to aquaculture is declining from 35.5 in 1950 to 1.47 in 2004 and the estimated figure for the corresponding ratio will be 0.51 in 2020. For clear visualization of the trend of aquaculture and capture production for the period from 1950 to 2004 is presented below (Figure 12.1).

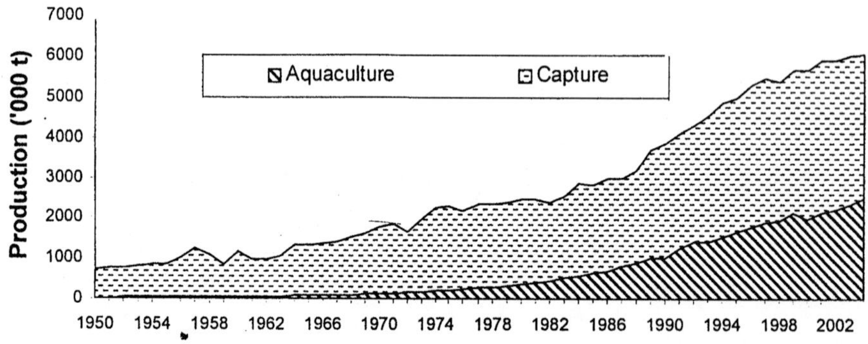

Source: Roy and Panda, 2007c

Figure 12.1: Total Fish Production in India (1950-2004)

The Figure 12.1 shows that the aquaculture production is growing at a faster rate compared to that of capture production that shows wide fluctuations between years.

Before proceeding for species-wise analysis of freshwater aquaculture fish production, it is thought of presenting the trend in fisheries in India for the period from 1950-2004 for total as well as Capture and Culture (Figure 12.2).

From the Figure 12.2, it is evident that the trend in total production has an upward growth during the period from 1950-2004 and the growth of capture fisheries is observed dwindling after 1971 whereas the trend of aquaculture production is observed steady and upword.

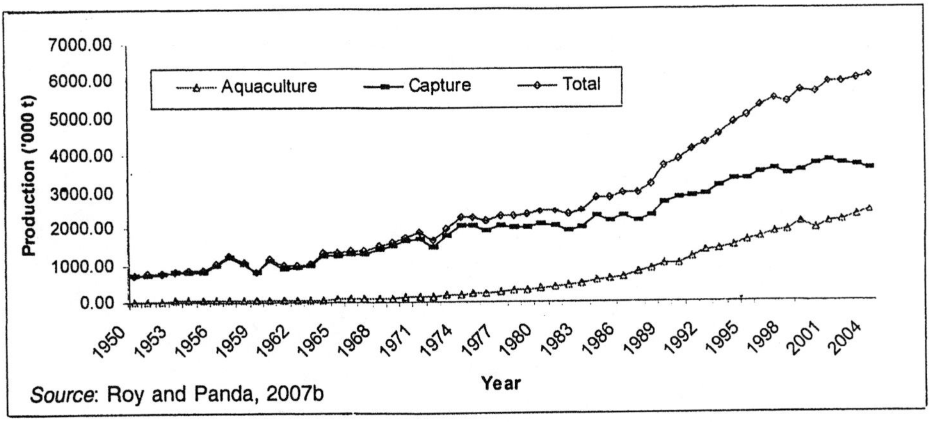

Figure 12.2: Trend in Capture, Culture and Total Fish Production

Aquaculture Production by Culture Environment

Aquaculture Production (decennial) based on different culture environment *viz.* freshwater aquaculture, brackish water aquaculture and marine aquaculture for the period 1950-2004 is given in the Table 12.2 along with annual percentage change is reproduced in Figure 12.3.

Table 12.2: Total Aquaculture Production in India based on Different Culture Environments (1950-2004)

Year	*1950*	*1960*	*1970*	*1980*	*1990*	*2000*	*2001*	*2002*	*2003*	*2004*
FW	17.91	44.84	121.65	361.31	982.14	1844.24	2015.66	2072.22	2197.09	2351.97
BW	0.00	0.00	0.02	3.87	35.00	96.72	102.93	114.97	115.88	120.37
MW	0.00	0.00	0.00	0.00	0.00	1.25	1.25	0.00	0.00	0.00
Total	17.91	44.84	121.67	365.18	1017.14	1942.20	2119.84	2187.19	2312.97	2472.34

From Figure 12.3 it is evident that annual percentage change in growth of freshwater aquaculture in almost all decades from 1950 to 2000 is comparatively significantly higher compared to those of brackish water and mariculture. It is also observed that in terms of total production freshwater aquaculture contributes about 95 per cent of total aquaculture production. Therefore, in the succeeding sections attention has been given on forecasting of freshwater aquaculture production by species.

Forecast of Species Wise Freshwater Aquaculture Production

Data on freshwater aquaculture fish production comprising various species like Catla (1950-2004), Climbing perch (1950-1999), Common carp (1993-2004), Grass carp (1993-2004), Mrigal (1950-2004), Rohu (1950-2004), Silver carp (1993-2004), Freshwater fishes nei (1950-2004), Giant river prawn (1989-2004), Monsoon river

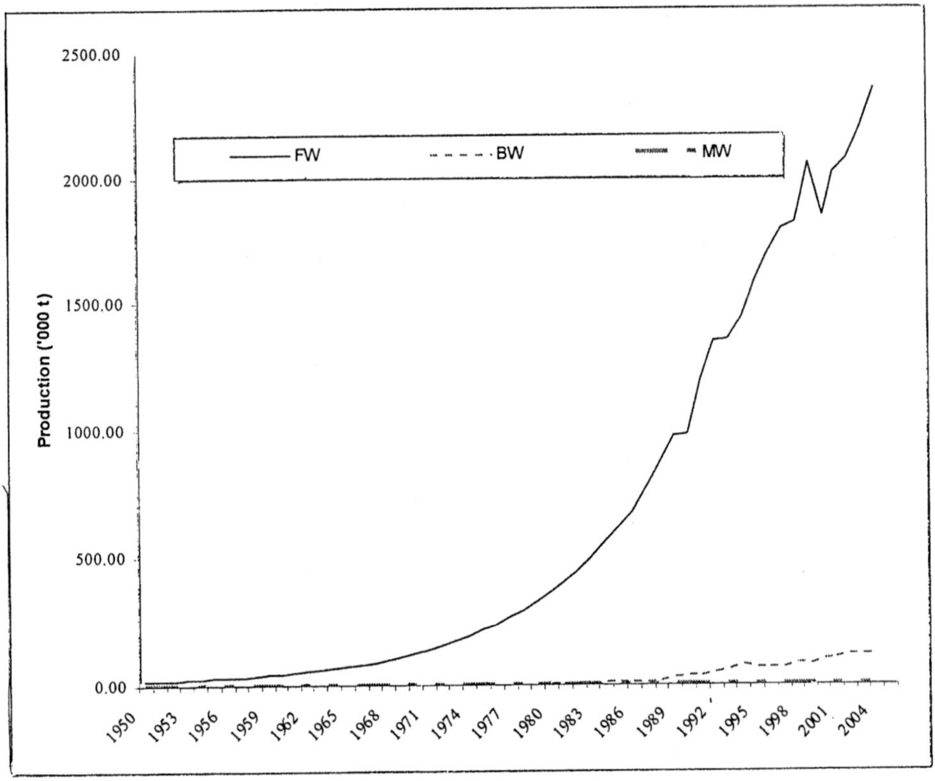

Figure 12.3: Total Aquaculture Production in India During 1950–2004 by Culture Environment

prawn (2003-2004), Snakeheads (Murrels) nei (1998-2004), Rainbow trout (1985-1992), Torpedo shaped catfishes nei (1968-2004), Kelee shad (2000-2004) were analysed. Species wise estimated production and coefficient of determination values for the period from 1950-2004 and forecasting values for the year 2020 were estimated using the regression model (Better Fit Method) available in the FISHSTAT Plus software package. MS-Excel was used for graphical presentation.

Aquaculture production in 2004 comprises mainly IMC-1.40 mmt (56.68 per cent), Exotic Carps-0.72 mmt (29.15 per cent), Catfish-0.04 mmt (1.62 per cent), Crustaceans-0.17 mmt (6.89 per cent) and other species-0.14 mmt (5.66 per cent). The IMC (1950-2004) constitutes Catla, Rohu and Mrigal, the Exotic (1993-2004) constitutes Silver Carps, Grass Carps and Common Carps, Catfish (1968-2004), Crustaceans (1970-2004) comprise Giant river prawn, Giant tiger prawn, Indian white prawn and Monsoon river prawn and Other species (1950-2004) constitutes Climbing perch, Rainbow trout, Freshwater fishes nei, Marine fishes nei, Kelee shad, Green mussel, Murrels etc.

Species wise actual production (2004), average production and estimated production (2020) of freshwater fish in India covering the period shown in brackets are reproduced below (Table 12.3).

Table 12.3: Species Wise Average Production, Actual Production (2004) Estimated Production (2020) and Coefficient of Determination Values of Freshwater Fish Production in India

Species	*Years (Count)*	*Actual Production (2004) (000' tons)*	*Average Production ('000 tons)*	*Estimated Production (2020) ('000 tons)*	*Coefficient of Determination (%)*
Catla (*C. catla*)	1950-2004 (55)	467.96	150	861	39.73
Climbing perch (*A.testudineus*)	1950-1999 (50)	65.00*	25	118	79.77
Common carp (*C. carpio*)	1993-2004 (12)	453.75	156	1318	76.94
Grass carp (*C.idella*)	1993-2004 (12)	100.64	68	366	15.13
Mrigal (*C. mrigala*)	1950-2004 (55)	442.78	125	814	39.61
Rohu labeo (*L. rohita*)	1950-2004 (55)	486.11	153	925	40.66
Silver carp (*H. molitrix*)	1993-2004 (12)	165.08	79	1434	11.36
Freshwater fishes nei	1950-2004 (55)	108.96	93	260	22.92
Giant river prawn (*M. rosenbergii*)	1989-2004 (15)	38.72	107	115	94.41
Monsoon river prawn	2003-2004 (2)	0.25	236	533	100.00
Snakeheads (=Murrels) nei	1998-2004 (7)	29.95	46	—	9.63
Rainbow trout	1985-1992 (8)	0.60**	472	—	2.49
Torpedo shaped catfishes nei	1968-2004 (37)	42.16	29	122	33.82
Kelee shad	2000-2004 (5)	0.18	13	871	53.47
Total (all species)	1950-2004 (55)	2402.14	949	6335	94.32

Source: Roy and Panda, 2007b.

Note: nei: Not Elsewhere Indicated; *: the figure for 1999 is taken as the figure for 2004; **: The figure for 1992 is taken as the figure for 2004.

Species Wise Contribution in Aquaculture Production (Present and Forecast)

The data on freshwater aquaculture fish in India available from the time series data of FishStat Plus comprise of Indian Major Carps (IMC) namely catla, rohu and mrigal; Exotic carps namely silver carp, grass carp and common carp; murrels, climbing perch, giant river prawn, monsoon river prawn, catfishes and other species. From the Table 12.4, it is observed that, the highest production was achieved in 2004 by rohu (486.11 th. tons) followed by catla (467.96 th. tons), common carp (453.75 th tons) and mrigal (442.78 th. tons). The total of actual production of all species is recorded as 2371.41 th. tons in 2004 whereas total estimated production of all species is forecasted as 6333.93 th. tons in 2020 which is 2.64 fold increase from the corresponding figure of 2004. The estimated production of prawn is projected by combining both giant river prawn and monsoon river prawn. The production of prawn in 2020 is estimated as 115.99 th. tons which are only 1.83 per cent of the total estimated freshwater fish production in 2020.

Table 12.4: Comparison of Contribution of Species Wise Production (Actual) in 2004 and (Estimated) in 2020 (000'tons)

Species	*Production (Actual) 2004*	*Contribution (%) 2004*	*Production (Estimated) 2020*	*Contribution (%) 2020*
Catla	467.96	19.73	861.15	13.60
Climbing perch*	65.00	2.74	118.04	1.86
Common carp	453.75	19.13	1,317.82	20.81
Freshwater fishes nei	108.96	4.59	260.36	4.11
Prawn (Monsoon river prawn + Giant river prawn)	38.96	1.64	115.99	1.83
Grass carp	100.64	4.24	365.79	5.78
Mrigal	442.78	18.67	814.02	12.85
Rohu	486.11	20.50	924.86	14.60
Silver carp	165.08	6.96	1,433.62	22.63
Catfishes	42.16	1.78	122.28	1.93
Total	2,371.41	100.00	6,333.93	100.00

Source: Roy and Panda, 2007b

N.B.: *: The figure for 1999 is taken as the figure for 2004.

For visualisation of contribution of various species in 2004 (actual) and 2020 (estimated) the values of per cent contribution is presented in Figure 12.4.

In order to visualize the future scenario in 2020, comparison of percentage contribution of different species of freshwater fish production in 2004 (actual) and 2020(estimated) is presented. An inspection of the Figure 12.4, it reveals that there will be remarkable increase in percentage contribution by silver carp with a highest

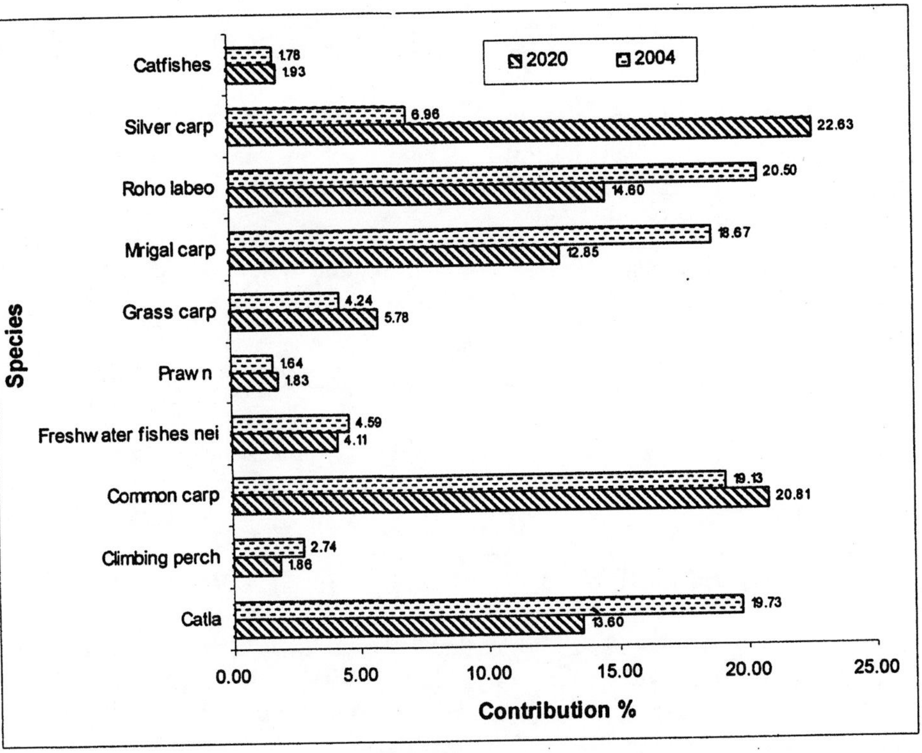

Figure 12.4: Comparison of Contribution (%) of Various Species in 2004 (Actual) and 2020 (Estimated)
***Source*: Roy and Panda, 2007b**

value of 22.63 per cent (2020) from 6.96 per cent (2004) followed by common carp and grass carp from 19.13 per cent (2004) to 20.81 per cent (2020) and from 4.24 per cent (2004) to 5.78 per cent (2020) respectively. It is significant to note that there will be marginal increase in percentage contribution by high value prawn from 1.64 per cent (2004) to 1.83 per cent (2020). There will be decrease in percentage contribution by each of the species of IMC like rohu (20.50 per cent to14.60 per cent), mrigal (18.67 per cent to 12.85 per cent) and catla (19.73 per cent to13.60 per cent) from 2004 to 2020. Similarly, there will be decrease in percentage contribution by climbing perch from 2.74 per cent (2004) to 1.86 per cent (2020).

Significant Observations

A significant finding of the present study is that the estimated figure for low valued silver carp is forecasted as 1434 th. tons in 2020 which is 22.63 per cent of the total freshwater fish production. The estimated production of high valued prawn is projected as 115.99 th tons which is 1.83 per cent of the total estimated freshwater fish production. High contribution of low value silver carp will help majority of low-income group of people in India to afford the price and on the other hand; contribution of high value prawn will add to the export earning of the country generating more revenue.

Intraspecies Comparison (Present and Forecast) of IMC

Further analysis of data indicating the intra species contribution of each of the species of IMC like catla, rohu and mrigal in 2004(actual) and 2020(estimated) is indicated in the table (Table 12.5).

Table 12.5: Comparison of Percentage Contribution of Rohu, Catla and Mrigal in IMC–2004 (Actual) and 2020 (Estimated)

Unit: (000'tons)

IMC	*Production (Actual) 2004*	*Contribution (%) 2004*	*Production (Estimated) 2020*	*Contribution (%) 2020*
Catla	467.96	33.50	861.15	33.12
Mrigal	442.78	31.70	814.02	31.31
Rohu	486.11	34.80	924.86	35.57
Total	1,396.86	100.00	2,600.04	100.00

Source: Roy and Panda, 2007b.

A close look at Table 12.5 indicates that there would be very marginal increase in percentage contribution by rohu from 34.80 per cent (2004) to 35.57 per cent (2020) and marginal decrease in percentage contribution both by catla (33.50 per cent to 33.12 per cent) and mrigal from (31.70 per cent to 31.31 per cent) in 2004(actual) and 2020 (estimated) respectively, which is visualized in the Figure 12.5.

Intraspecies Comparison (Present and Forecast) of Exotic Carps

Species wise comparison of contribution of Common carp, Grass carp and Silver carp in 2004 (actual) and in 2020 (estimated) is presented in Table 12.6 and Figure 12.6 for clear visualization.

Table 12.6: Comparison of Contribution (Per cent) of Various Species in Exotic Carps 2004 (Actual) and 2020 (Estimated)

Unit: (000'tons)

Exotic Carps	*Production (Actual) 2004*	*Contribution (%) 2004*	*Production (Estimated) 2020*	*Contribution (%) 2020*
Common carp	453.75	63.07	1,317.82	42.28
Grass carp	100.64	13.99	365.79	11.73
Silver carp	165.08	22.95	1,433.62	45.99
Total	719.47	100.00	3,117.23	100.00

Source: Roy and Panda, 2007b.

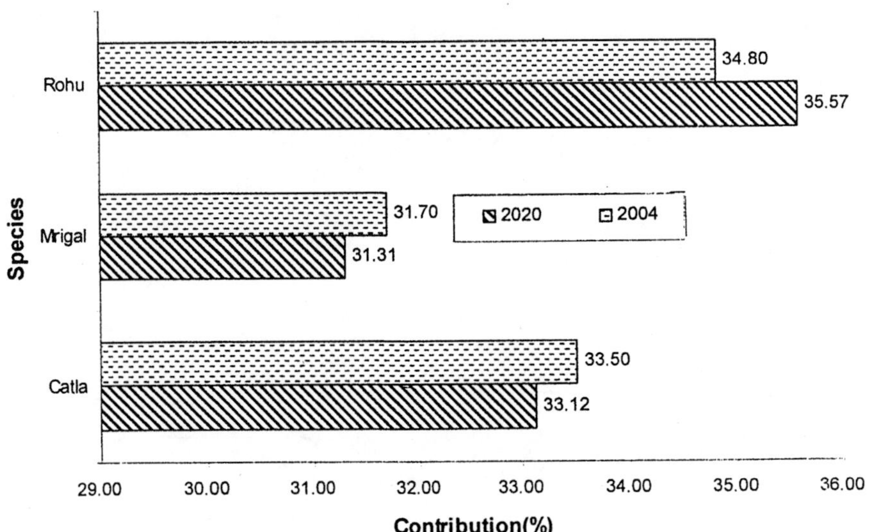

Source: Roy and Panda, 2007b

Figure 12.5: Comparison of Contribution (%) of Rohu, Catla and Mrigal in IMC 2004 (Actual) and 2020 (Estimated)

From the Figure 12.6, it is observed that contribution of common carp will be decrease significantly from 63.07 per cent to 42.28 per cent and grass carp from 13.99 per cent to 11.73 per cent in 2004 (actual) and 2020 (estimated) respectively. A reverse trend is observed in case of silver carp that shows that there will be almost 9 (nine) fold increase in production from 165.08 th. tons in 2004(actual) to 1433.62 th.tons in 2020 (estimated) contributing 45.99 per cent of total exotic carps in 2020 compared to that of 22.95 per cent recorded in 2004.

Comparison of Production (Present and Forecast) between IMC and Exotic Carps

The Table 12.7 shows the data for intergroup comparison of percentage contribution by IMC and Exotic carps to total freshwater fish production comprising of all species. From the table, it is found that production of IMC and Exotic carps in 2004 were recorded as 1396.86 th.tons (58.15 per cent) and 719.47 th. tons (29.95 per cent) respectively where as the corresponding estimated production in 2020 is projected as 2600.04 th.tons (41.04 per cent) and 3117.23 th. tons (49.21 per cent) respectively. Hence, it is indicated from the study that, the forecast figure of the Exotic groups in 2020 will be higher than that of IMC. The comparison is visualized in the following figure (Figure 12.7).

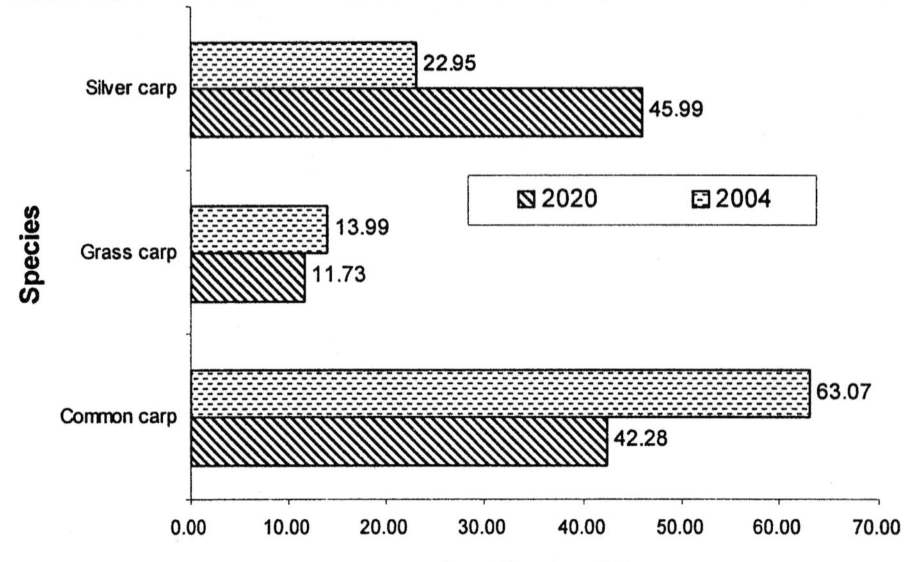

Source: Roy and Panda, 2007b

Figure 12.6: Comparison of Contribution (%) of Various Species in Exotic carps (2004 and 2020)

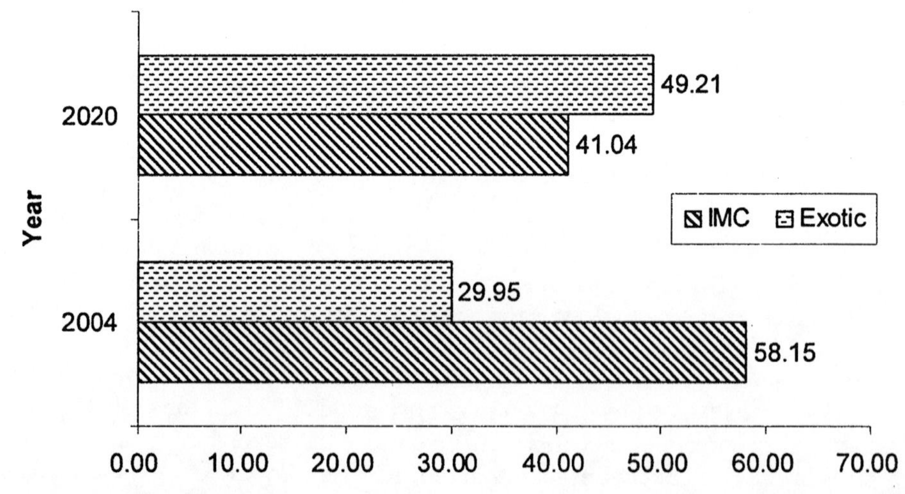

Source: Roy and Panda, 2007b

Figure 12.7: Comparison of Contribution (%) of IMC and Exotic Carps

Table 12.7: Comparison of Production of IMC and Exotic carps in 2004 (Actual) and 2020 (Estimated)

Unit: (000'tons)

Groups	Production (Actual) 2004	Contribution (%) 2004	Production (Estimated) 2020	Contribution (%) 2020
IMC	1396.86	58.15	2600.04	41.04
Exotic	719.47	29.95	3117.23	49.21
Total production of all species	2402.14		6335.00	

Source: Roy and Panda, 2007b.

Test of Consistency in Reported Production

The Table 12.8 shows basic statistics of production for different species of freshwater aquaculture. The coefficient of variation (CV) value is found to be highest in case of silver carp (154) followed by common carp (118) and grass carp (76). The low CV values for catla (14), rohu (14) and mrigal (13) indicate the consistency in production of these species over the years compared to those of exotic carps.

Table 12.8: Basic Statistics for Different Species of Freshwater Aquaculture Fish

Unit: (000'tons)

	Maximum	Minimum	Mean	S.D.	C.V.
Catla	510000	340100	433526	58635	14
Rohu	519321	342500	445854	64186	14
Mrigal	480000	332600	413877	52281	13
Total	1509321	1015200	1293257	174813	14
Common carp	453747	4894	156435	183876	118
Grass carp	137200	10585	67981	51473	76
Silver carp	390907	1505	78512	121159	154

Source: Roy and Panda, 2007b.

Test of Goodness of Fit of Model

Forecast values are estimated based on regression model 'better fit method' of Fishstat plus, FAO. Accordingly, the values of co-efficient of determination for various species were obtained that range from murrels (9.6 per cent) to giant river prawn (94.4 per cent). Further observation exclusively on IMC reveals that co-efficient of determination (R^2) values for catla, rohu and mrigal are 39.7 per cent, 40.6 per cent and 39.6 per cent respectively. R^2 values for exotic carps namely silver carp, grass carp and common carp are also found out as 11.36 per cent, 15.1 per cent and 76.9 per cent respectively (Table 12.3). So it may be concluded that variations explained by a single independent variable (time) in case of rohu, catla, mrigal are of moderate range and in case of giant river prawn and common carp, variations explained are highly satisfactory. Interestingly, R^2 value for total fish production in found to be 94.32 per cent that is highly satisfactory.

Forecast of Group-wise Freshwater Aquaculture Fish Production

Further analytical study on various species/group of species estimates of production of freshwater aquaculture reveals that contribution of IMC in the year 2020 is 2.6 mmt (44.52 per cent), Exotic carps 1.55 mmt (26.54 per cent), Catfish 0.12 mmt (2.05 per cent), Crustaceans 0.94 mmt (16.10 per cent) and other species 0.23 mmt (3.94 per cent). The graphical presentation showing the comparison between actual production (2004) and the forecast production (2020) for IMC, Exotic carps, catfish, crustaceans and other species is demonstrated below (Figures 12.8–12.12).

Comparison of Contribution (Present and Forecast) for Various Groups of Freshwater Aquaculture Fish Production

From the Figure 12.13, it is clear that contribution of high value crustaceans will enhance to the extent of 6.5 per cent with higher foreign exchange value. If the present trend continues contribution of exotic species of carps will be enhanced to the extent of 15 percent at the cost of IMC whose contribution will be reduced from 56 per cent in 2004 to 37 per cent in 2020. The impact of this change may have serious consequences on socio-economic conditions of farmers in terms of consumption and nutritional requirement of vast majority of farmers thriving on aquaculture countrywide.

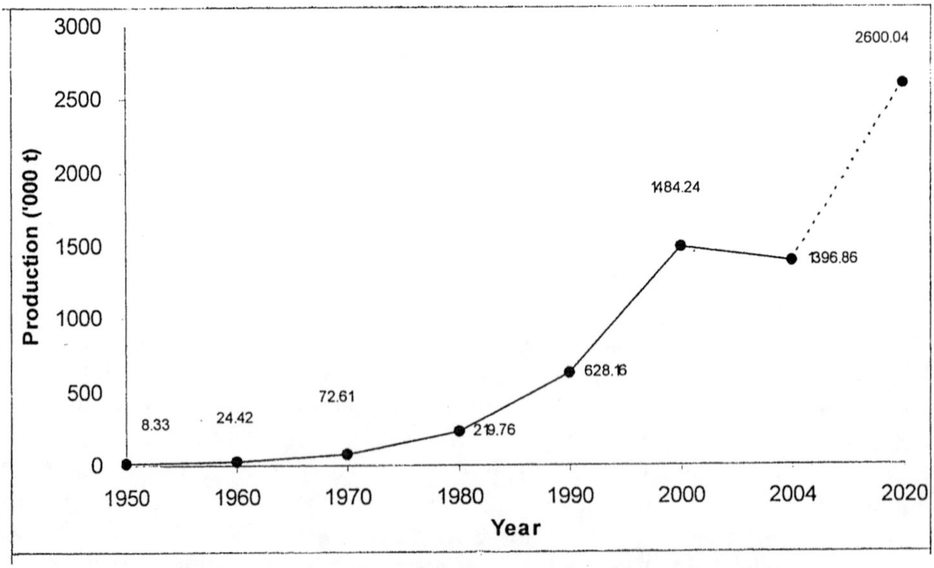

Figure 12.8: Trend in Production of IMC (1950–2004) and Forecast (2020)

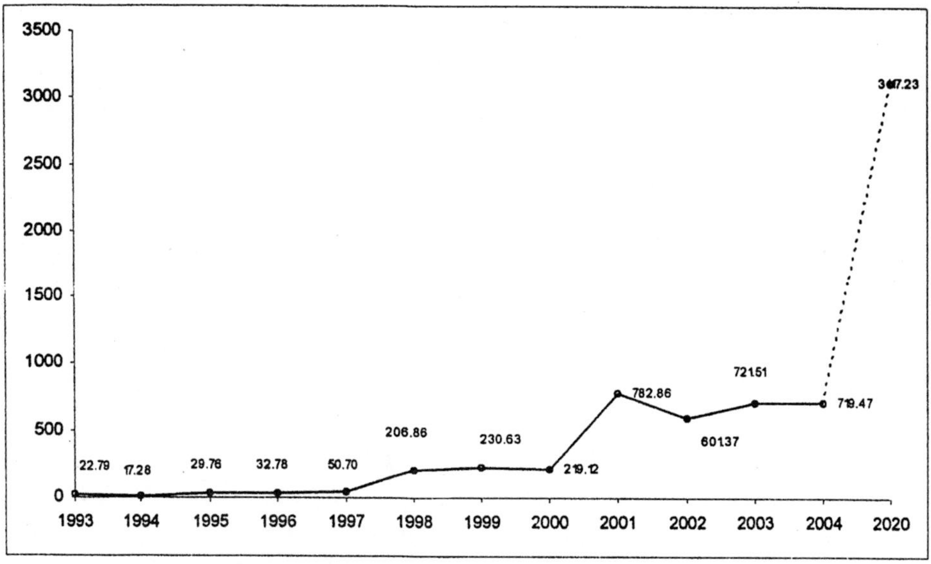

Figure 12.9: Trend in Production of Exotic Carps (1993–2004) and Forecast (2020)

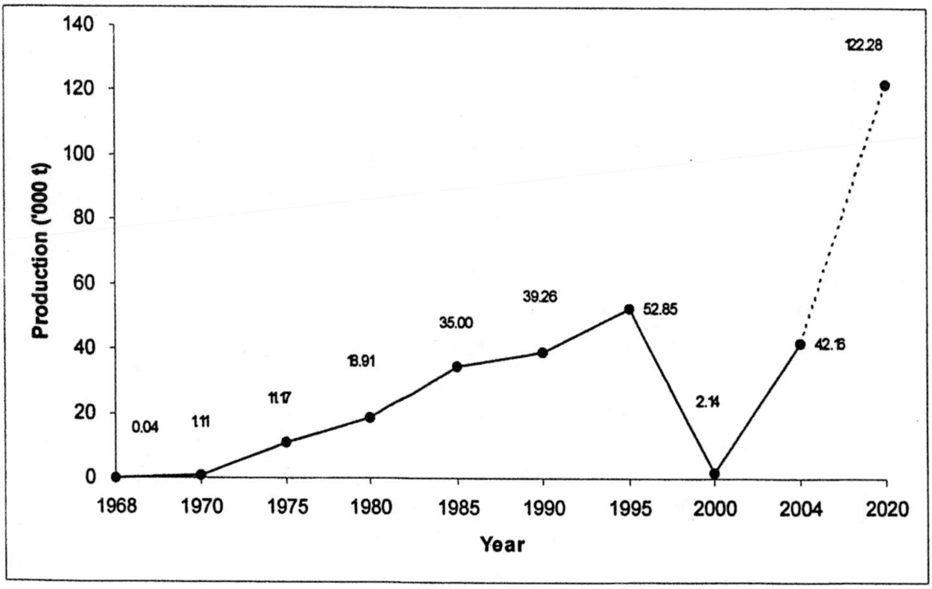

Figure 12.10: Trend in Production of Catfish (1968–2004) and Forecast (2020)

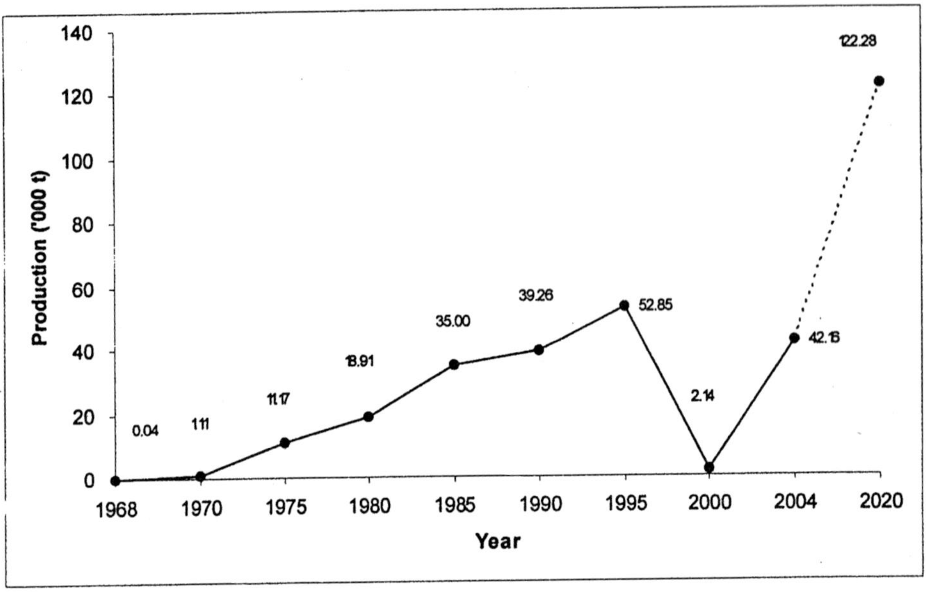

Figure 12.11: Trend in Production of Crustaceans (1970–2004) and Forecast (2020)

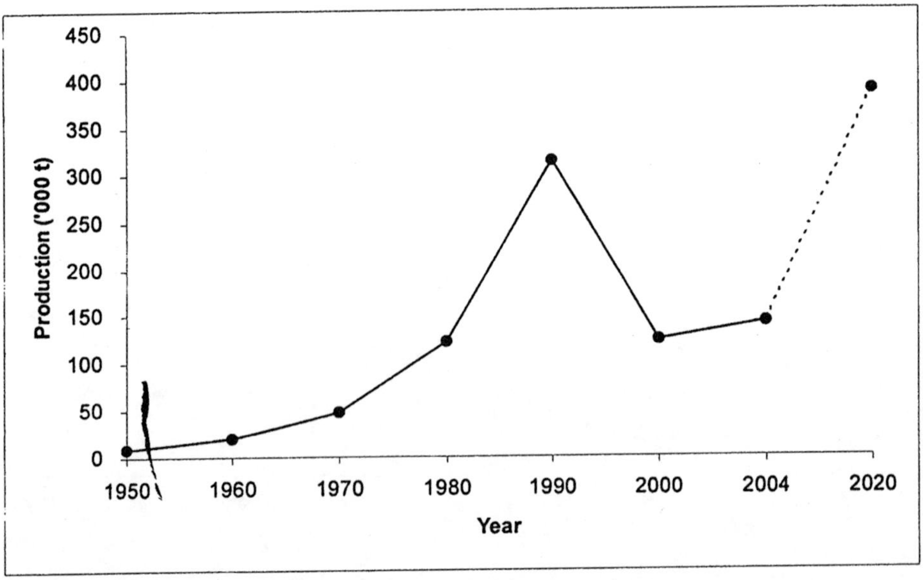

Figure 12.12: Trend in Production of Others (1950–2004) and Forecast (2020)

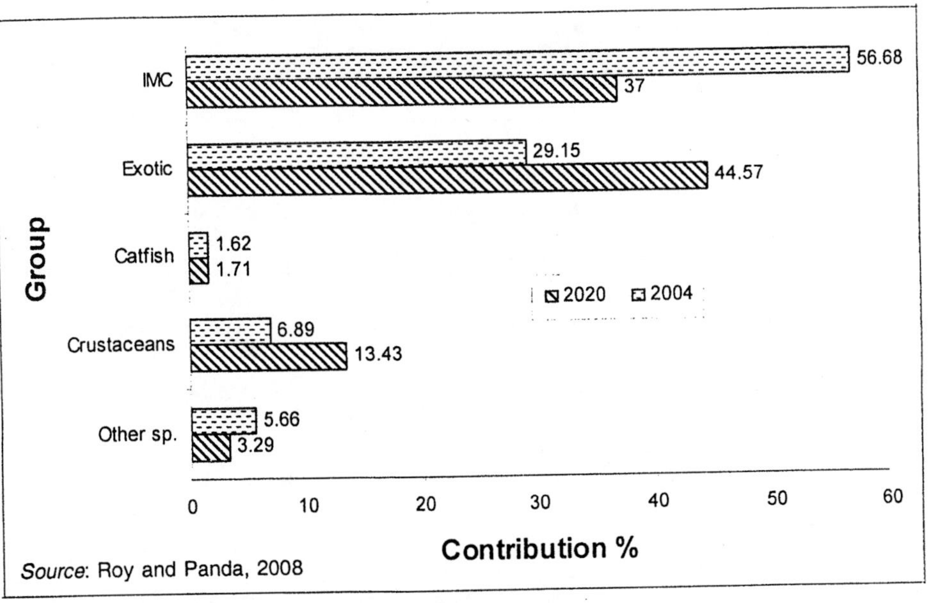

Figure 12.13: Group Wise Comparison of Freshwater Aquaculture Fish Production in India in 2004 (Actual) and in 2020 (Estimated)

Capture Fisheries Production in India

Group-wise Time Series Data

The time series data on capture fisheries in India for the period from 1950-2004 was downloaded from FishStat Plus, FAO (version 2.30) and compiled group wise as per the definitions and terminology described by the FAO (FAO, 2007). In order to categorise various groups, options for four groups namely ISSCAAP Group, FAOSTAT Group, Main Grouping and Custom Group are mentioned in the software of FishStat plus, FAO. Out of these four groups, the FAOSTAT group was found to be conventional and therefore adopted for compilation of data. According to FAOSTAT group, capture fisheries in India comprises various major groups like cephalopods (1957-2004), crustaceans (1950-2004), demersal marine fish (1950-2004), freshwater and diadromous fishes (1950-2004), marine fish nei (1954-2004), molluscs (excl. cephalopods) (1981-2004) and pelagic marine fish (1950-2004) are considered for the present study. The species wise composition of each group is mentioned in details in the Table 12.9.

Coefficient of determination values for each group for the existing period were determined and forecasting production for the year 2020 were estimated using the regression model (Better Fit Method) available in the FISHSTAT Plus, FAO. MS-Excel was used for various computational work and graphical presentation of species wise fish composition.

Table 12.9: Major Groups Showing Species Under Each Category as per Grouping by FAOSTAT

Group	*Species*
Cephalopods	cephalopods nei
Crustaceans	antarctic krill, freshwater crustaceans nei, giant tiger prawn, marine crabs nei, marine crustaceans nei, natantian decapods nei, penaeus shrimps nei;
Demersal marine fish	bombay duck, croakers, drums nei, flatfishes nei, goatfishes, hairtails, scabbardfishes nei, lantern fish, lizardfishes nei, mullets nei, percoids nei, pike congers nei, ponyfishes (=slipmouths) nei, sea catfishes nei; sharks, rays, skates, etc. threadfins, tasselfishes nei, unicorn cod;
Freshwater and diadromous fishes	cyprinids nei, diadromous clupeoids nei, freshwater fishes nei, freshwater siluroids nei, kelee shad, snakeheads (=murrels) nei
Marine fish nei	marine fishes nei
Molluscs	freshwater molluscs nei and marine molluscs nei
Pelagic marine fish	Anchovies, etc. nei, barracudas nei, bigeye tuna, butterfishes, pomfrets nei; carangids nei, clupeoids nei, clupeoids nei, false trevally, flying fishes nei, flying fishes nei, frigate and bullet tunas, halfbeaks nei, Indian mackerel, Indian oil sardine, Indo-Pacific king mackerel, Indo-Pacific sailfish, Jacks, crevalles nei, kawakawa, longtail tuna, marlins, sailfishes,etc. nei, narrow-barred spanish mackerel, Pompanos nei, seerfishes nei, skipjack tuna, streaked seerfish, swordfish, tuna-like fishes nei, wahoo, wolf-herrings nei, yellowfin tuna, anchovies, etc. nei, barracudas nei, bigeye tuna, butterfishes, pomfrets nei and carangids nei.

Source: FAOSTAT, 2004; *Note*: nei: Not elsewhere indicated.

Group wise Present (2004) and Estimated Production (2020) for Various Groups of Capture Fishery

From the Table 12.10 it is evident that the production of marine fish showed 163.1 times increase (1954-2004) followed by cephalopods and molluscs that demonstrate 125.1 times (1957-2004) and 52.8 times (1981-2004) increase over the initial period respectively. The forecasted production for 2020 predicts that the increase of molluscs will be 7.58 times followed by crustaceans 1.66 times whereas increase in production of marine fish will be only 0.3 times over the period 2004. An analysis of coefficient of determination values for different groups reveals that in case of crustaceans, demersal marine fish and pelagic marine fish the values are determined as 87.93, 83.66 and 69.02 per cent respectively justifying that the regression model (better fit method) used in the software of Fish STAT Plus, FAO is a valid and good estimator. In case of marine fish nei the coefficient of determination value (3.92) is the least among all the groups, which is a result of high fluctuation in the production values of marine fish nei for the period 1954-2004.

Table 12.10: Group Wise Production (2004) and Estimated Production (2020)

Groups	*Years (Count)*	*Prodn. in 2004 (mmt)*	*Times Increase in 2004 Over Initial Period*	*Estimated Prodn. in 2020 (mmt)*	*Times Increase Over 2004*	*Coeff. of Determination*
Cephalopods	1957-2004 (48)	0.08	125.10	0.03	0.38	12.55
Crustaceans	1950-2004 (55)	0.46	6.63	0.77	1.66	87.93
Demersal marine fish	1950-2004 (55)	0.85	3.40	0.54	0.64	83.66
Freshwater and diadromous fishes	1950-2004 (55)	0.73	3.81	0.36	0.50	16.57
Marine fish nei	1954-2004 (51)	0.78	163.08	0.22	0.28	3.92
Molluscs (excl. cephalopods)	1981-2004 (24)	0.003	52.84	0.02	7.58	33.16
Pelagic marine fish	1950-2004 (55)	0.71	3.55	0.53	0.74	69.02
Total		3.613		2.47		

Source: Roy and Panda, 2008.

Intra Group Comparison of Present and Forecast Production from Capture Fisheries

In order to compare production and relative contribution, group wise production figures along with contribution is given in Table 12.11.

Table 12.11: Comparison of Various Groups (Per cent Contribution) in 2004 and 2020 (Estimated)

Groups	*Production in 2004 (mmt)*	*% Contribution (2004)*	*Estimated Production in 2020 (mmt)*	*% Contribution (2020)*
Cephalopods	0.08	2.08	0.03	1.16
Crustaceans	0.46	12.84	0.77	31.22
Demersal marine fish	0.85	23.42	0.54	21.85
Freshwater and diadromous fishes	0.73	20.24	0.36	14.73
Marine fish nei	0.78	21.65	0.22	8.77
Molluscs (excl.cephalopods)	0.003	0.07	0.02	0.81
Pelagic marine fish	0.71	19.71	0.53	21.46
Total of all species	3.61	100.00	2.47	100.00

Source: Roy and Panda, 2008.

For easy comprehension the figures are expressed in terms of contribution of each group of fishes for the actual (2004) and estimated (2020) is presented in Figure 12.14.

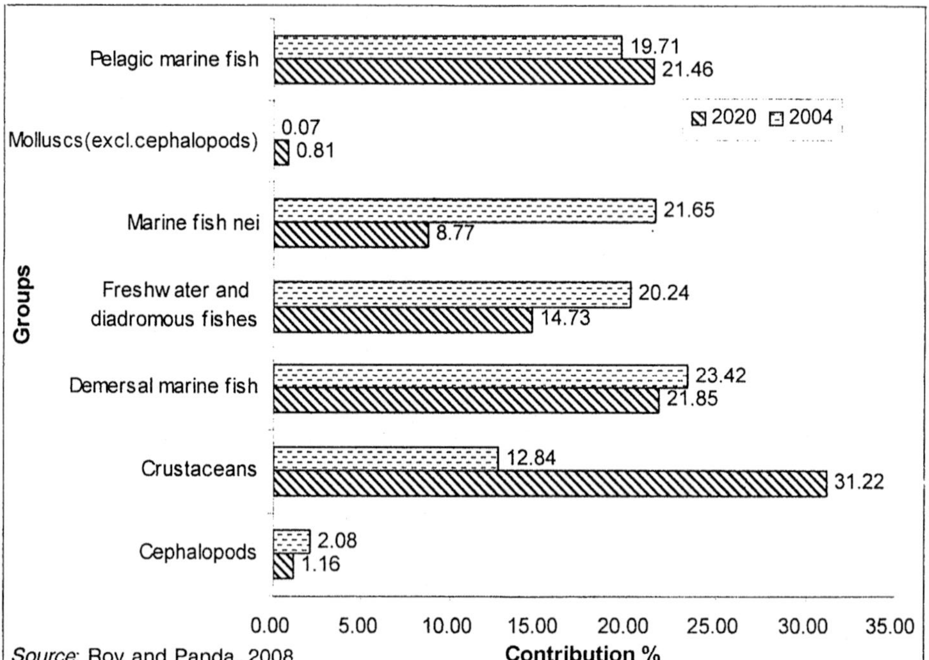

Source: Roy and Panda, 2008

Figure 12.14: Comparison of Contribution of Various Groups in 2004 Actual and in 2020 (Forecast)

From Figure 12.14, it is observed that in 2004 demersal marine fish, marine fish nei, freshwater and diadromous fishes, pelagic marine fish and crustaceans contribute 23.42, 21.65, 20.24, 19.71 and 12.84 per cent of total capture production respectively. But the estimated production figures based on present trends indicate that crustacean, demersal marine fish, pelagic marine fish, freshwater and diadromous fishes and marine fish nei will contribute to the extent of 31.22, 21.85, 21.46, 14.73 and 8.77 per cent respectively in 2020. It is worth mentioning that in 2020 the contribution by high value crustaceans will increase significantly. This is likely to generate higher revenue in foreign exchange. The figures reveals the fact that contribution of crustaceans will be about two and half times more in 2020 compared to the present contribution of 12.84 per cent (2004). Contribution of Pelagic marine fish will increase marginally about 2 per cent more in 2020 compared to that in 2004 and molluscs from 0.07 to 0.81 per cent. On the other hand, during the same period, marine fish nei will decrease from 21.65 to 8.77 per cent, freshwater and diadromous fishes from 20.24 to 14.73 per cent, demersal marine fish from 23.42 to 21.85 per cent and cephalopod from 2.08 to 1.16 per cent. Group wise trend of capture fishing was drawn based on data for the period 1981–2004 for visualisation (Figure 12.15).

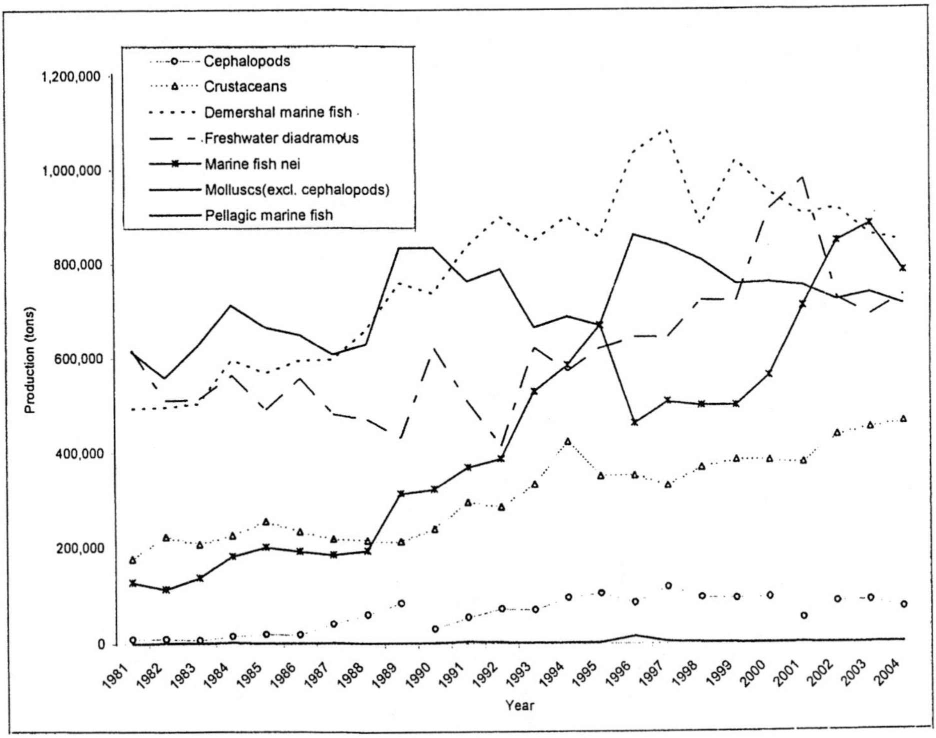

Figure 12.15: Growth of Capture Production (Group-wise) During the Period 1981–2004

Trend in Growth of Group Wise Capture Fish Production

From Figure 12.15, it is clear that marine fish nei demonstrated highly fluctuating behaviour after 1996 and then reached to the maximum in 2003 with the maximum production of 0.88 mmt. followed by a declining trend during 2003-04.

Trend of Ratio of Capture : Culture and Culture : Capture

Trends of ratio of productions of capture to culture and culture to capture during the period 1950 to 2004 are presented in Figure 12.16. It appears from the figure that over the years, the ratio of production by capture to culture is declining at a steady rate during last 20 years and in case of culture to capture the trend in ratio is progressing at a greater rate during the period from 1972 to 2004. These trends are reflected in forecast figure of both capture and culture production of 2020.

Socio-Economic Impact of Likely Decline in Capture Fisheries

In India thousands of fishermen living in about 4000 villages are engaged in activities related to capture fisheries spread all over the coastal line of about 8118 km; continental shelf of 5.3 lakh sq. km.; covering 1896 landing centres and many major and minor fishing harvours and ports. Besides many are engaged in motorized boats,

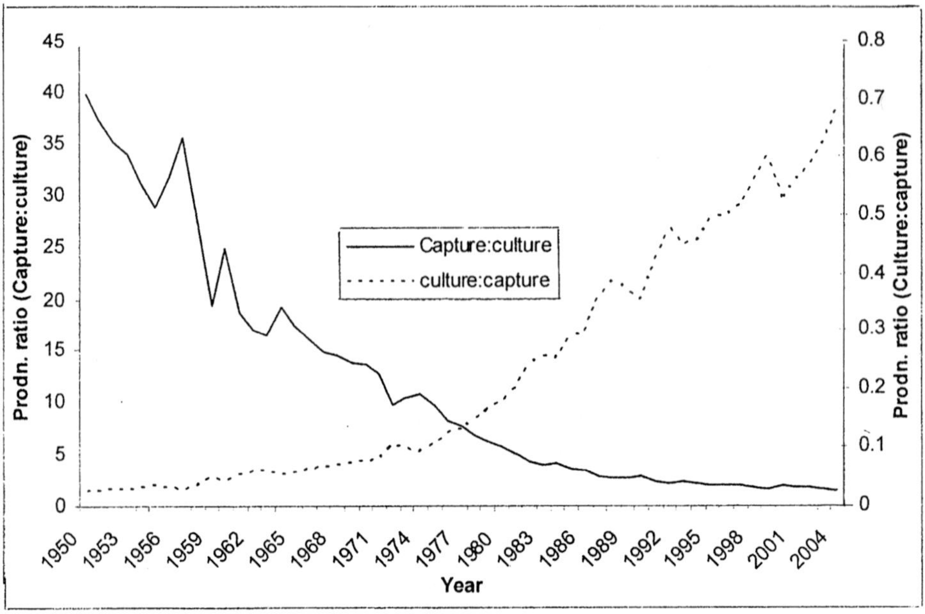

Figure 12.16: Trends of Ratio (Capture : Culture and Culture : Capture) of Fish Production During 1950–2004 in India

net making, packaging and transportation, ice plants etc. Livelihood of millions of rural people is dependent on capture fisheries. If there is a reduction of production in 2020 to the extent of 31.6 per cent as predicted, many will be out of job resulting in nutritional food scarcity, social unrest and loss of revenue for government.

References

Anon, 2006a. Pamphlet, National Fisheries Development Board. Department of Animal Husbandry, Dairying and Fisheries, Ministry of Agriculture, Govt. of India, Block 401-402, Maitri Vihar, Huda Commercial Complex, Ameerpet, Hyderabad-500038, Andhra Pradesh, India.

Chandran and Prajneshu. 2005. Nonparametric regression with autocorrelated errors methodology for describing India's marine fish production data. *Indian Journal of Fisheries.* Vol. 52, no. 2, pp. 151-158.

Cremer, MC, Baoxin, Z, Schmittou, HR, Jian, Zhang. 1998. Status and forecast for the freshwater aquaculture production and feed industries in China, *Aquaculture 98 Book of Abstract.* p. 124.

Delgado, Christopher L.; Nikolas Wada, Mark W. Rosegrant, Siet Meijer and Mahfuzuddin Ahmed; Sustainable solutions for ending hunger and poverty;2020 Vision Food Policy Report; IFPRI and World Fish Centre. www.apps.fao.org/.

FAO, 2005. Aquaculture production, 2003. Yearbook of fishery statistics–Vol.96/2. Food and Agriculture Organization of the United Nations, Rome, Italy.

Global Aquaculture Production, FAO, 2007; Collection Global Aquaculture Production, Overview(1); http://www.fao.org/fi/website/FIRetrieveAction.do

Gopakumar, K. 2003. Indian aquaculture. Journal of Applied Aquaculture, 13(1/2):1-10.

Kumar, P, Dey, Madan M. and Paraguas, Ferdinand J., 2005. Demand for Fish by species in India: Three stage Budgeting Framework. Agricultural Economic Research Review, Vol.18, pp.167-186.

Vacek, J. 1973. The importance and role of forecasting activity in fisheries, *Bul. VURH Vodn. Vyzk, Ustav Ryb. Hydrobiol.* Vol. 9. *no.3*, pp.3-8.

Walia, SS, Jain, RC. 1998. Non-linear statistical model for pre-harvest forecasting of fish production from inland ponds. *Indian Journal of Fisheries.* Vol. 45, no. 1, pp. 75-78

Roy, A. K. and Panda N. 2007a. Forecasting of Fish Production in India in 2020. Workshop on Sustainability of Indian Aquaculture Industry (Sustain-Aqua 07) held during 28-29 Sept., 2007 at IIT, Kharagpur.

Roy, A. K. and Panda N. 2007b. Projection of Freshwater Aquaculture Fish Production in India. In: Fisheries and Aquaculture strategic outlook for Asia, Book of Abstracts, 8th Asian Fisheries Forum (8aff) organized by Asian Fisheries Society, Indian Branch, November 20-23, 2007 at Cochi, India.

Roy, A.K. and Panda, N. 2007c. Forecasting Capture Fisheries Production in India. National Symposium on Ecosystem Health and Fish for Tomorrow conducted by IFSI and CIFRI at Barrackpore during 14-16, December, 2007.

Alam, M.F., 2002. Socioeconomic aspects of carp production and consumption in Bangladesh. *In*: D.J. Penman, M.G. Hussain, B.J. Mcandrew and M.A. Mazid (Eds.) Proceedings of a workshop on Genetic Management and Improvement Strategies for Exotic Carps in Asia, 12-14 February, 2002, Dhaka, Bangladesh. Bangladesh Fisheries Management Institute, Mymensingh, Bangladesh.

Delgado, Christopher L.; Nikolas Wada, Mark W. Rosegrant, Siet Meijer and Mahfuzuddin Ahmed; Sustainable solutions for ending hunger and poverty; 2020 Vision Food Policy Report; IFPRI and World Fish Centre www.apps.fao.org/.

De Silva, S. 2001. A global perspective of aquaculture in the new millennium. *In*: Subasinghe, R.P., Bueno, P., Phillips, M.J., Hough, C., McGladdery, S.E. and Arthur, J.R. (Eds.) Aquaculture in the Next Millennium. Technical Proceedings of the Conference on Aquaculture in the Third Millennium, Bangkok, Thailand, 20-25 February, 2000, NACA, Bangkok and FAO, Rome, pp.431-459.

Delgado, C.L., Wada, N., Rosegrant, M.W., Meijer, S. and Ahmed, M. 2003. Fish to 2020. Supply and Demand in Changing Markets. International Food Policy Research Institute, Washington, D.C. and World Fish Center, Penang.

FAOSTAT, 2004. Fisheries Data Available online : http://faostat.external.fao.org/faostat/collections?subset=fisheries. Last updated in February 2004. Rome, FAO.

Tacon, Albert J., 2007. Aquaculture Production Trends Analysis: FAO

Fish Stat Plus. 2004. Aquaculture Production: Quantities 1950-2002. Updated 30 April 2004. Rome, FAO.

Josupeit, H. and Franz. N. 2004. Aquaculture–Trade, trends, standards and outlooks. Globefish. Fisheries Department, Rome, FAO. Available online: http://www.globefish.org/index.php.

Wijkstrom, U.N. 2003. Short and long-term prospects for consumption of fish. Veterinary Research Communications, 27 suppl. 1:461-468.

Wijkstrom, U.N. and New, M.B. 1989. Fish for feed: a help or a hindrance to aquaculture in 2000? INFOFISH International, 6/89:48-52.

Ye, Y. 1999. Historical consumption and future demand for fish and fishery products: Exploratory calculations for the years 2015-2030. FAO Fisheries Circular No.946, Rome, FAO.

http://www.fao.org/spfs

Modelling, Forecasting, Artificial Neural Network and Expert System in Fisheries
Editors: Ajit Kumar Roy & Niranjan Sarangi
Published by: DAYA PUBLISHING HOUSE (2009)

Pages 179–192

Chapter 13

Expert Systems and Simulation Studies in Agriculture/Aquaculture

J. Panduranga Rao

Introduction

Expert System is one of the important application oriented branches of artificial intelligence. In the past decade, a great deal of expert systems and simulation models had been developed and applied to many fields such as office automation, science, medicine and aquaculture including agriculture. At the beginning of development of the agriculture expert systems, the areas selected are applications to diagnosis of diseases and pests of various crops. In recent years, development of the expert system in the fields of agriculture domain has been paid much attention by many countries, including developed and also developing countries. The complexity of problems confronting farmers like yield losses, soil erosion, diminishing market prices due to international competition, increasing chemical pesticide costs and economic barriers hindering adoption of farming strategies necessitates that they become expert managers of all aspects of their farming operations. The agricultural researchers need to address problems of farm management and discover new management strategies to improve farm production. Numerical methods have failed because understanding about crop systems are qualitative based on experience and cannot be mathematically represented. Expert Systems are Computer programs that are different from conventional computer programs as they solve problems by mimicking human reasoning processes, relying on logic, belief, rules of thumb, opinion and experience. The most commonly known type of knowledge based system is the rule based expert system in which the experience and knowledge of a human expert is captured in the form of IF-THEN rules and facts which are used to solve problems. In general terms main objective of the expert system is to generate expert advice for use by a decision-maker.

Expert systems are used to aid:

- ☆ Single point decisions. *e.g.* Planning
- ☆ Designing. *e.g.* Design of an irrigation system
- ☆ Selection. *e.g.* The most suitable Crop variety or market outlet
- ☆ Diagnosis or identification. *e.g.* of a livestock disorder
- ☆ Interpretation. *e.g.* of a set of financial accounts
- ☆ Prediction. *e.g.* of extreme events such as thunderstorms and frost;
- ☆ A sequence of tactical decisions throughout a production cycle. *e.g.* plant protection and nutrition decisions, livestock feeding.

Further, an expert system for a particular decision problem can be used:

As a stand alone advisory system for the specific knowledge domain perhaps with monitoring by a human expert:

- ☆ To provide decision support for a high-level human expert
- ☆ To allow a high-level expert to be replaced by a subordinate expert aided by the expert system
- ☆ As a delivery system for extension information
- ☆ To provide management education for decision makers
- ☆ For dissemination of up-to-date scientific information in a readily accessible and easily understood form, to agricultural researchers and advisers.

An Agricultural Extension Expert System is a Decision Support System for Agricultural Extension Agents who have to decide what advice to be offered to farmers who have to decide what action to take are based on it. It is one of the most efficient extension tool to take the technology from scientists to the farmers directly without any dilution of content which normally creeps in because of the number of agencies involved in normal technology transfer systems.

Components of an Expert System

The inference engine (software) is that part of the program which contains reasoning capability. It interacts with a knowledge base (IF...THEN...ELSE statements), which contains information about how to solve problems within the problem domain. Juxtaposed to this is the global memory where the knowledge based system records information relating to a specific problem that it is trying to solve. Much of this information comes from the user but the inference engine to record its own conclusions and to remember its chain of reasoning also uses the memory. By comparing what it knows about the problem domain in general with what it knows about the specific problem, the inference engine tries to proceed logically towards a better solution.

The principal power of the Expert System is therefore derived from the knowledge that the system possesses. Use of Expert Systems are most appropriate where the expert possesses complex knowledge about a highly specific subject area for which no complete algorithms have been formulated and no complete theory exists. They have become commercially important because they provide a way of achieving expertise and making it available when it is needed. They are particularly valuable where existing experts are expensive or not available.

At present in most of the developed countries complex medical, mechanical design and agricultural extension problems are being solved by expert systems. In these countries expert systems are freely accessible to individual farmers at their farm offices, where they can get subject matter expert opinion on their field specific problems from the expert systems by hooking to their agricultural network which in turn connected with all the agricultural universities and research stations.

Advantages of the Expert systems

- ☆ Provide a framework with which to capture and apply non-algorithmic knowledge.
- ☆ Have the ability to imitate human thought and reasoning
- ☆ Make modification of knowledge very convenient
- ☆ Makes interaction more user friendly due to ability of interpretation and transparence
- ☆ With the machine learning technique knowledge can be acquired automatically and directly from experimental data and real time examples
- ☆ Provide expert level recommendations in a language, which is understandable to users
- ☆ Have the ability to handle uncertain information.

From the researchers point of view, knowledge based systems have a potential to help in organizing and synthesizing knowledge and information of different types. It is possible to focus and apply diverse avenues of research to solve difficult problems and link together quantitative data, simulation models, basic research results into knowledge based model.

The expert system is shifting the focus of the research community to knowledge dissemination in contrast to knowledge accumulation. The expert system in combination with powerful personal computers and new devices like CD-ROM has the potential to open whole warehouses of accumulated knowledge to agricultural produce.

The Development of Expert Systems

Expert systems developed in the 1960s and 1970s were typically written on a mainframe computer in the programming language based on list Processing (LISP). Evolving from university research laboratories, they were limited to the applications developed by these research sites. Most of these expert systems were not intended for commercial use. They incorporated the specific knowledge of the experts. About the problem area, termed "domain knowledge." Problem-solving heuristics (or "rules of thumb") and inferences capabilities, and an interface mechanism between the user and the system. Some examples of these systems include MACSYMA, developed at the Massachusetts Institute of Technology (MIT), for assisting individuals in solving complex mathematical problems: Standford University's MYCIN, which diagnosed bacterium and meningitis infections, which is the first diagnostic Expert System ever developed and the University of Pittsburgh's INTERNIST/CADUCEUS, which aided internal medicine diagnosis and decision making. These scientists created a general-purpose tool for developing expert systems now called a "shell".

Expert System Shells

It was not until the 1980 are that commercial shells were introduced for a variety of classes of computers. Some of the more popular shells today are Ist-Class, ADS, ART, CRYSTAL.XSYS, Gold works, Guru, Level 5, Nexpert object, KDS, KES, MI, personal Consultant S.I, TIMM, and VP Expert. These packages not only provide the necessary tools for developing an expert system such as the user-system interface, interface mechanism, rule editor, and code optimizer, but can generally be run on microcomputer in addition to minicomputers and mainframes, are reasonably priced, and provide powerful features without requiring an individual to learn the mechanics of an Artificial Intelligence (AI) language or purchase specialized hardware. With the development and widespread use of shells, the range of expert systems applications had tremendously increased. Their introduction had played a major role in expanding expert system applications into areas such as aquaculture, agriculture, management, finance, office automation, computer selection and networking, legal processes, manufacturing, equipment training, personnel training, education, transportation, oil and geology, science and medicine.

Some of the Expert Systems and Simulation models developed in Aquaculture have been discussed here.

Fastfish-on Farm Assessment of Stress Level in Fish

The main goal of the FASTFISH project (2006-2009) is to develop a database and expert system for on farm welfare and stress level assessment in farmed fish at different ontogenetic stages (FAST-TOOL). This will include validated non-invasive behavioural stress level indicators that can be used to monitor and quantify acute and chronic stress levels. The main focus areas for potential indicators will be feeding motivation and behavioural control mechanisms.

Decision Support for Pond Aquaculture Planning and Management

Decision-making for pond aquaculture requires knowledge of the effects of management practices on fish performance, soil and water quality. It is also necessary to examine economic and environmental consequences of various practices. A decision support system POND which enables definition of an entire aquaculture facility (in terms of location, ponds, fish populations, and species), and provides analysis capabilities in the form of simulation models. POND models can be used to examine the implications of management practices such as feeding, fertilization, POND models can be used to examine the implications of management practices such as feeding, fertilization, POND models can be used to examine the implications of management practices such as feeding, fertilization, liming, stocking and water exchange rates on facility-level fish production, liming, stocking and water exchange rates on facility-level fish production.

Web-based Expert Systems: Benefits and Challenges

Convergence of technologies in the Internet and the field of expert systems have offered new ways of sharing and distributing knowledge. However, there has been a general lack of research in the area of web-based expert systems (ES). This paper addresses the issues associated with the design, development, and use of web-based ES from a standpoint of the benefits and challenges of developing and using them. The original theory and concepts in conventional ES were reviewed and a knowledge engineering framework for developing them was revisited. The study considered three web-based ES: WITS-advisor–for e-business strategy development, Fish-Expert-for fish disease diagnosis.

A Simulation Model of the Response of Harp Seals to Alternative Harvesting Strategies

The current management regime for harp seals is based on periodic updates of replacement yield, defined as that level of annual catch, which stabilizes the population from one year to the next. This strategy is appropriate only when the regulated population has reached the long-term management objective. In the case of harp seals, there appears to be no long-term management objective with respect to desired population levels. In this report a variety of alternative harvesting strategies are considered and the population response of harp seals to these strategies are simulated using two population models. (George H. Winters and Daniel S. Miller, Focus Technologies Inc. 1).

Significant findings of the simulation studies are as follows :

The replacement yield for 1999 is 402,000 animals, if recent mortality and reproductive rates hold.

(*a*) the maximum sustainable yield (MSY) is estimated to be about 280,000 seals at an MSY population level of about 3.5 to 4.0 million animals.

(*b*) a one-time cull of 2.0 million animals in the year 2000 would reduce the population to the approximate level of the MSY population, after which a long-term harvest of 275,000 animals could be sustained under normal environmental conditions.

(*c*) alternative harvest levels such as a kill of 400,000 seals will gradually reduce the current population to the MSY level over a period of several decades.

(*d*) in the absence of a commercial hunt the harp seal population will gradually increase to a maximal level of 7.1 million animals. Regulation of the harp seal population at the MSY population level should be considered as a management option for a variety of biological and conservation benefits such as optimizing harp seal reproductive and growth rates, improving overall ecosystem yield and increasing the probability of recovery of depressed ground fish populations.'

Bayesian Estimation of Genetic Distance Between Populations

Shuichi Kitada (Department of Aquatic Bi-sciences, Tokyo University of Fisheries, Minato, Japan; kitada@tokyo-u-fish.ac.jp): Genetic effects of hatchery releases on wild populations have become matters of concern in stock enhancement activities. Genetic differences between hatchery and wild populations have been examined by allozyme and DNA analyses on several target species. A homogeneity test of genotype frequencies was employed. However, testing null effects is not meaningful if the power of the test is not considered. The essential thing is to evaluate the genetic distance of hatchery-reared and wild populations, from which appropriateness of the releases can be considered. Here a Bayesian estimation proçedure of genetic distance was proposed. Dirichlet priors of the genotype frequencies of two populations are specified. The posterior distribution of the genetic distance is then obtained by Monte Carlo simulation. To avoid sensitivity to the subjective priors, a hierarchical model was proposed. The procedure makes it possible to evaluate the genetic effects statistically, taking full account of information content in the data. As a case study, data on red sea bream are analyzed.

Comparison of Otolith Shape Using Fourier Series Analysis Between Species, Among Populations, and Among Stocks of Perch (*Perca fluviatilis* and *P. flavescens*).

Stephen C. Chong (Department of Biology, Queens University, Kingston, Ontario chongs@biology.queensu.ca): There is a growing body of research comparing otolith shape characteristics in different fish species. Otolith shape is a morphological characteristic that is often species specific. The majority of studies have concentrated on intraspecific otolith shape comparisons as a possible management tool for discriminating among stocks of fish. Growth rate has been implicated as a major factor affecting otolith shape among fish stocks. However very few studies have compared the relative importance of genetics vs. growth rate in determining otolith shape. In this study otolith shapes among populations of Eurasian (*P. fluviatilis*) and yellow perch (*P. flavescens*) were compared using Fourier series analysis. By comparing the otolith shapes among populations of two closely related percid species, the influence of genetics vs. growth rate was evaluated. The data collected demonstrated that the otolith shape difference between Eurasian and yellow perch was substantial, providing further evidence that these two species should be considered separate. Furthermore, otolith shape comparisons among allopatric and sympatric perch stocks suggest that the growth rate influence is more important when genetic differences are far below the level considered for species-level separation. When a high degree of genetic differentiation exists, growth rate has less influence on otolith shape.

Uncertainty Projection in Fishery Models: Some Examples Using Interval Analysis and Fuzzy Arithmetic

Saul B. Saila(University of Rhode Island, Graduate School of Oceanography, Narragansett, ssaila@gso.uri.edu):It is increasingly recognized that uncertainties are ubiquitous in fishery science. Major forms of uncertainty include random fluctuations, uncertainty in the estimates of model parameters, and structural uncertainty, which reflects our lack of knowledge of the behavior of fishery systems. Uncertainty from random fluctuations and parameter estimates has been addressed by several methodologies, including Bayesian analysis, bootstrapping, jackknifing, and likelihood methods. This report addresses parameter estimation error and its propagation through simple models. Interval analysis and fuzzy arithmetic are introduced and applied to the models. The first simple application relates to the bio-accumulation of a toxicant in fish tissue. Next, uncertain parameters are applied to the Beverton-Holt and Ehrhardt and Ault mortality estimation models using spiny dogfish Squalus acanthias as the source of inputs. A spreadsheet model of relative yield per recruit and yield per recruit is then developed and examined for spiny dogfish using uncertain parameters. The results obtained from these examples suggest that interval analysis and fuzzy arithmetic may be suitable methods for dealing with uncertainty when the assumptions of traditional probabilistic error analysis are not met.

Choosing Complexity in Stock-Production Models

Michael H. Prager (NMFS Southeast Fisheries Science Center), Mike.Prager@noaa.gov: Using computer simulation, two common production models were compared: the simpler logistic model descended from the work of Pearl, Lotka, and Schaefer; and the more complex generalization of it from the work of Pella and Tomlinson. The latter includes among its parameters an exponent that is difficult to estimate precisely. The two models have similar properties and were fit by similar methods, conditioned on observed catches. It has been examined how well each model estimates the underlying parameters of simulated populations following one or the other model, but with random error. The study provides some insight into the mean–variance tradeoffs involved in choosing between these two models.

Analysis of Minimum Size Limit for the Red King Crab Fishery in Bristol Bay, Alaska

Gordon H. Kruse (Alaska Department of Fish and Game, Division of Commercial Fisheries, gordonk@fishgame.state.ak.us):Low stock status of red king crabs in Bristol Bay prompted an industry initiative to reduce the minimum size limit for the commercial fishery from 165 mm (6 ½ in) to 152 mm (6 in) carapace width. The rationale is to reduce non-target handling mortality, a suspected contributor to depressed stocks. Reduced size limit diminishes sublegal by catch, increases legal catch rate, and reduces fishing effort needed to attain total allowable catch (TAC) levels. Average TAC (in terms of weight) declines owing to smaller mean size of landed crabs because harvest rate is defined in terms of crab numbers. Size limit has little effect on yield per recruit because of reduced catchability of small crabs.

Cumulative catch under the reduced size limit would exceed cumulative catch under the current size limit after two decades of implementation. Net benefits of the reduced size limit are larger and accure more quickly if handling mortality rates exceed 20 per cent, while the reduced size limit never yields positive economic benefits if handling mortality rates are less than 10 per cent.

Can Multivariate Analysis be Useful in Forecasting Recruitment in the Chesapeake Bay

Robert J. Wood (Virginia Institute of Marine Science, College of William and Mary Gloucester Point, rjw@vims.edu):Understanding the recruitment dynamics of estuarine and marine fish stocks is a primary goal in fisheries science. Developing multispecies approaches to fisheries management is another. Perhaps surprisingly, progress in both of these areas may be facilitated by examining recruitment variability in a multispecies context. Is recruitment of species correlated over space and time? This paper investigates the multivariate structure within five multispecies, fishery-independent recruitment data sets available for the Chesapeake and Delaware Bay estuaries. Principal components analysis (PCA) reveals that the strongest signal within each of these data sets describes an alternating dominant-subordinate relationship between coastally spawned species and those species that are anadromous or semi-anadromous. This relationship has existed for at least the 40 year period of record (since 1958). The strength and persistence of this pattern across time, space, data set collection methods, and species suggests that the negative relationship between these species groups is real and may be the result of synoptic scale climatic forcing.

RHABSIM Version 2.1
Thomas R.Payneand Associates
890 L Street, Arcata, California USA 95521

RHABSIM (Riverine Habitat Simulation) is a fully integrated program for river hydraulics and aquatic habitat modelling using the Instream Flow Incremental Methodology (IFIM). Running in Microsoft Windows and DOS, it is an extensive conversion of the PHABSIM hydraulic and habitat simulation system developed by the US Fish and Wildlife Service.

Knechtel, Charles Douglas

cknechte@cresnet.org

Application of mathematical modelling, statistical analysis, computerized simulation, and computer programming to solve fisheries, environmental, and/or other problems were discussed.

Some Agricultural Expert Systems

A number of research and development projects in progress world wide, are applying Expert system technology in various areas. An attempt is made here to discuss some models of Expert systems in the field of Agriculture in India and abroad.

Indian Expert Systems

Rice-Crop Doctor

National Institute of Agricultural Extension Management (MANAGE) has developed an expert system to diagnose pests and diseases for rice crop and suggest preventive/curative measures. The rice crop doctor illustrates the use of expert systems broadly in the area of rice production through development of a prototype, taking into consideration a few major pests and diseases and some deficiency problems limiting rice yield.

This prototype is a result of joint effort by the experts from NIIT and computer professionals of MANAGE while the subject matter expert knowledge on rice pathology and entomology, has been obtained from Scientists of Andhra Pradesh Agricultural University (APAU) and Directorate of Rice Research (DRR).

A number of diseases and pests have been included in the system for identification and suggesting preventive and curative measures. The diseases included rice blast, brown spots, sheath blight, rice tungro virus, false smut fungi, bacterial leaf blight, sheath rot and zinc deficiency disease. The pests include stem borers rice gall midge, brown plant hopper, rice leaf folder, green leaf hopper and Gundhi bug.

The Logic Flow

The brief logic flow of the above expert system is as follows:

1. The part of the plant where symptoms have been the observed is given by the extension officer
2. The basic symptoms are given as input;
3. Considering these symptoms the user is expected to give further information based on other visual symptoms
4. At this step the disease and pest are identified
5. The user is then given the option to either stop or further diagnose other disease/pest or get preventive or curative measures on these.

All these operations are done using a "mouse", which helps a new user to get acquainted with the system faster and operate it easily. In this system, computer, graphics, digitized pictures and text are being used following a "Decision Tree", logic.

e-Sagu: IT-based Personalized Agricultural Extension

To improve agricultural productivity and profitability for the farmer, research is going on International Institute of Information Technology to develop e-Sagu ("Sagu" means cultivation in Telugu language) which is personalized and cost-effective agricultural extension system to deliver personalized agricultural expert advice to each farm in a timely manner by exploiting advances in Information Technology. In e-Sagu, the agricultural experts generate the expert advice based on the crop situation received in the form of both text and digital photographs.

Expert Systems on Horticultural Crops

The first software for use by the grape cultivators was prepared by the Indian Institute of Horticultural Research Institute, Bangalore. This spontaneous response made them to undertake similar software for providing guidance to mushroom cultivators, which became extremely popular and a large number of growers using it regularly for getting solutions to their problems. The Institute has launched into an effort to give a comprehensive package of practices of about 148 horticulture crops for cultivation in the 4 Southern states of Kerala, Tamil Nadu, Karnataka and Andhra Pradesh.

AGREX

Center for Informatics Research and Advancement, Kerala has prepared an Expert System called AGREX to help the Agricultural field personnel give timely and correct advice to the farmers. These Expert Systems find extensive use in the areas of fertilizer application, crop protection, irrigation scheduling, diagnosis of diseases in paddy and post harvest technology of fruits and vegetables.

Farm Advisory System

Punjab Agricultural University, Ludhiana, has developed the Farm Advisory System to support agri-business management. The conversation between the system and the user is arranged in such a way that the system asks all the questions from user one by one which it needs to give recommendations on the topic of farm Management. The inputs are encouraging and acceptance by farmers is very good.

Exploring Artificial Intelligence Techniques for Selection of Geographical Zones for Sericulture

Central Silk Board, Bangalore has developed an Expert System for selection of Geographical Zones for Sericulture. Higher Yields in Sericulture can be obtained if suitable Zones for its practice can be identified based on parameters like rainfall, rainy days, relative humidity fluctuations and temperature etc. Expert Systems will help to identify the suitable to raise mulberry and rear healthy silk worms.

Computer Modelling for Irrigation Scheduling, Crop Growth and Estimation: A Case of Ground Nut Crop in Middle Gujarat

Mr. Veerparia and Mr. S.M.Patel have developed a simulation model using relationship between air temperature and soil temperature up to 30-cm depth, which predict soil temperature at different depths based on only maximum and minimum air temperature. Soil temperature variations at different depth of soil influence crop growth. A biological process, nutrient information and nutrient availability is controlled by soil temperature and soil water. Soil temperature has a profound influence on seed germination, root and shoot growth nutrient uptake and crop growth. Even a difference of one degree centigrade in soil temperature can have a significant effect on plant growth. For summer groundnut crop in Gujarat soil temperature of above 18 degree centigrade at 10 cm. depth is required for proper seed germination. If soil temperature at below 18 degree centigrade, it is not advisable to grow for sowing the groundnut on that day. The models can be applied for sowing of crops like mustard, millets etc.

Computer Assisted Agriculture through Distributed Knowledge Based Expert System

There are three levels at which the basic process of agriculture is happening. They are 1. Low level farmers. 2. Middle level agricultural officers. 3. Higher level research institutions. Each level requires a data depending upon its requirements with interrelationships among them. The farmers interface is designed in such a way that the communication between the expert system and the farmer will be in the farmer's own native language. The knowledge base acts as a bridge between farmers and research institutions. The production systems at farmers level knowledge base which are not able to find an answer are formed as unanswered dynamic frames and tried for solution by agricultural officers. These frames are transferred to research institutions with the same additional view of points of agricultural officers. Hence research institutions can come to know about the new undiscovered problems that exist at farmers level.

Expert Systems of Other Countries

Knowledge representation and expert system tools have been applied successfully into agriculture in China, such as adequate fertilization, plant protection cultivation water resource management and irrigation, raising and management of livestock and poultry. The Expert Systems have pushed the growth of agriculture and also the research of Artificial Intelligence in China. They have produced evident economic effect and social impact.

Expert Systems for Cotton Crop Management

This Expert System has been developed by the US Department of Agriculture to provide appropriate management recommendations to cotton growers.

CALEX

This is a blackboard based integrated expert decision support system for agricultural management, developed at University of California. CALEX can be used by growers, pest control advisors, consultants and other managers.

VARIEX

This expert system developed at Technical University of Brno, Czechoslovakia enables section of the best cultivators for different agricultural situations.

Weiping Jin Expert System

In this System various processes like fertilizing, irrigating, spraying insecticide etc process rely on crop state analysis. However, there is a lot of fuzzy phenomenon in crop state analysis. It is therefore necessary to analyze or to measure these fuzzy phenomena by means of fuzzy set theory for crop management. An expert system that can deal with fuzziness has been developed for crop management in PR. China. This paper shows the principal construction of the expert system, CMES, for growers to manage their crop. CMES provides support for crop growth control system (CGCS);

i.e., advises growers about optimal population and structure of crop in planting stage and when and what measures to adopt on their crop to keep at optimal state, to avoid Infestation in various stage of growth and development of crop, and finally to obtain high crop productivity at harvest stage.

LEY Expert Systems

A RF-telemetry based computer-controlled, automated, remote, real-time weather data acquisition and reporting system is described in Washington State. Cooperation among Washington State University the National Weather Service and the U.S. Bureau of Reclamation and several private grower organizations have made this system possible. Data is collected, processed and transmitted to the NWS hourly. These hourly updates of actual conditions are broadcast on the NOAA weather band during the spring frost season to assist fruit growers with frost protection. Real-time weather data are also being used in applications such as irrigation scheduling, crop protection and pest management.

A decision support system by P.H.Heinemann exists as a prototype and is being refined to assist managers with the development, implementation, and management of frost protection systems. The program is divided into four main modules: frost protection strategies, operational management, forecasting, and instrumentation. It is being designed to be an addition to other commodity management expert systems that are commercially available or are under development.

Using a unique satellite downlink facility at by R.R. Getz Auburn University, real time meteorological data are being processed on a network of advanced computer workstations. This data feeds in a series of computer models that generate site specific predictions of temperature, dew point, wet bulb temperature, and other parameters are used to alert Alabama fruit, vegetable, and nursery growers about freeze condition. Model outputs are distributed to Extension specialists and others over a campus-wide fiber optic communications system and to all county Extension offices over the ACENET computer network. During critical freeze situations, live information can be disseminated via satellite to media and rural satellite dish owners using the Auburn University uplink facility. A pilot network of automated weather stations incorporating a 10-m tower for monitoring inversions has been installed to enhance the program.

John R. Barrett "The logic and terminology of researchers are not usually that of producers!"

The capabilities that knowledge engineering offers for solving ill-structured agricultural problems where data, information and knowledge are incomplete has excited many agricultural scientists. Limited acceptance has resulted, primarily because of lack of understanding of the decision- making process, inadequate user involvements, and improper problem definitions. User behavior during decision-making is beginning to be realistically considered in program development. From another perspective, most programs represent interests and knowledge from a scientific viewpoint. These do not satisfactorily address the needs of the decision-making users, expressed in familiar words, terms, symbols and logic. Efforts to include client-users in the development of software to improve the management of production with emphasis on corn and soyabeans are introduced. A procedure and results of observing expert producers to identify missing or incomplete information, to determine critical success factors and to define logic leading to decisions made by expert producers over managerial time are presented.

G I S Expert Systems by Naiqian Zhang

The expert system software WHEATWIZ was developed in 1987 as an effective tool to assist Kansas farmers, extension workers, and agri-business personnel in variety selection for hard red winter wheats (Shroyer *et al.*, 1987). The software offered three modes of operation.

The concept of "prescription farming" or precision application of agronomic inputs may dramatically increase the attention, a farmer gives to the spatial distribution of nutrients in the soil and methods of fertilizer application. By using a geographic information system (GIS) and a cotton model, GOSSYN, together as a spatial crop simulation, a methodology for investigating the implications of incorporating prescription farming programs for cotton production is investigated. The data contained in the GIS, the result of intensively sampling a cotton field for nitrate information, is used as the primary input to the model. Hydrologic profile information is also mapped using the GIS and used as input to the model. The model is run for each unique combination of inputs. Crop response is mapped back the GIS for spatial analysis. Information pertinent to prescription farming programs which can be extracted from the spatial simulation is discussed. Precision application of agronomic inputs may necessitate increased soil sampling and increased complexity of the technology on tractors and fertilizer applicators. However, this may also help to increase crop yields, as well as to provide a degree of environmental protection from excessive use of chemicals.

CLIPS Expert System by L. L.M. Rhykerd

A prototype alfalfa (Medicago sativa L.) management CLIPS (C Language Integrated Production System) expert system has been developed by Purdue University agricultural scientists. This form of artificial intelligence provides an extension tool, which will enable farmers to very economically use a computer program to reach conclusions concerning profitable alfalfa production that normally would require consultation with a forage expert. To date, the management considerations and the sequence in which they were built into the knowledge base are: soil drainage, soil pH, soil P test, soil K test, use of alfalfa crop, chemical weed control, expected longevity of stand, variety recommendation, method and rate of seeding and pure live seed. The knowledge base of these CLIPS expert system can easily be updated, as new information becomes available or can be modified for use in other states or farming regions.

The Promise of Agriculture Expert System

Expert Systems will play a major role in the dissemination and application of useful knowledge leading to economic growth and higher standards of living. They are not only the vehicles to apply experts knowledge to particular problems but are potentially powerful learning resources to help end users to develop their own expertise. The capacity building potential may be one of the most important aspects of the spread of the expert systems in the developing countries. For both developed and developing countries this can bring more productivity and employment in agriculture through wider and more diverse applications of new scientific results. It also increases the scope for individual managerial initiative by all farmers reinforcing local abilities to solve local problems.

Conclusion

The Expert system must be developed in local languages, which will help the farmers to develop their own expertise, which in turn will enhance the production and productivity of agriculture. These expert systems must be available in village booths, which act as information resource center for the farmers in the villages.

References

Artificial Intelligence and authorities and Government, Proceedings of the conference on Artificial Intelligence Industry and Govt., Nov 23-25, 1989 Hyderabad, India.

Artificial Intelligence Technology Applications and Management Proceedings of the 4th International Computing Congress, Dec 15-18,1993, Hyderabad.

Conference on Information for Sustainable Agricultural development Abstracts: NIC and Ministry of Agriculture, 1994.

Expert System in Developing Countries Practice and promise. Edited by Charles K Mann and Stephen R.Ruth, 1992.

I T challenges and opportunities proceedings of 30th Annual Convention of the Computer Society of India, Sep 1995.

Knowledge based systems in Agriculture. Richard E. Plant and Nicholas D. Stave. McGraw-Hill 1991.

Proceeding IFAC Workshop an Expert System and Agriculture. International Academic Publishers, Beijing 1992.

Modelling, Forecasting, Artificial Neural Network and Expert System in Fisheries
Editors: Ajit Kumar Roy & Niranjan Sarangi
Published by: DAYA PUBLISHING HOUSE (2009)

Pages 193–205

Chapter 14

Recent Trends in Artificial Neural Network (ANN) and Fuzzy Logic and their Application in Aquaculture

M. Balakrishnan and R.C. Srivastava

Introduction

Artificial neural networks (ANN), inspired by the structure and processes of biological cognition and learning, were developed initially to model biological functions. They learn from experience and can rapidly solve difficult computational problems. In the past decade, research into ANN has shown explosive growth. ANN models have often been applied in chemistry and in physics, for example in studies of speech and image recognition. Most applications of ANN in biology have concerned medicine and molecular biology (Lo *et al.*, 1995) [1] but with a few examples in the Ecological and environmental sciences beginning in the 1990s. For instance, Colasanti (1991) [2] perceived similarities between ANN and ecosystem structure and functioning and recommended the utilisation of this tool in ecological modelling. In a review of computer-aided research in biodiversity, Edwards and Morse (1995) [3] also emphasised the potential importance of ANN. Other examples can be found in different fields of ecology, such as modelling of the greenhouse effect, predicting phytoplankton production (Scardi 1996) [4], and predicting various parameters in fish ecology (Brosse *et al.*, 1999b,c) [5] [6]. Most of these studies demonstrated that ANN performed better than classical linear and nonlinear modelling methods such as multiple linear regressions or generalised additive models. However, very few studies have dealt with the application of ANN to the ecology of stream invertebrates or of multi-species assemblages (Schleiter *et al.*, 1999) [7]. In this paper we assess the ability of ANN to predict ecological parameters relating to freshwater invertebrates and fish. We provide a general Description of the ANN approach and then test the capacity of ANN models to predict the diversity of macroinvertebrates in tributaries.

The application of artificial intelligence (AI) in agricultural and natural resources management started with the development of expert systems for problem solving and decision-making [8]. ANN is an attempt to mimic the computational process of the brains that consists of a number of interconnected processing element (or) "neurons" [9]. Usually at one end of the network data fed in and at the other end, the network produces outputs. In between there are layers of neurons that may or may not are depending on their current state, their inputs their earlier layers and their own properties [10]. An interest in the use of artificial neural networks (ANN), are distributed and parallel computational algorithm, which have been efficiently used, in various practical problems. ANN has also begun to emerge as an alternative approach for modelling nonlinear and complex phenomena in Fisheries science. However artificial neural network (ANN) is an aspect of computational intelligence research that has been recently shown efficiency in describing non-linear biological systems [10].

Features of ANN

There are different neural network models developed since the first prototype neural network was proposed in 1943 [11]. Neuron science was influential in development of neural networks, but physiologists and engineers also contributed of neural network systems- [12] created a great deal of interest when he designed and developed the perception. One of the most widely used neural networks in yield prediction is the back-propagation feed-forward network (also referred as the Multi-Layer Perceptions (MLP) networks are widely used due to their simplicity and excellent performance [13]. As one of many possible examples, this section provides a brief introduction to the major features of the BPN and MLP.

Artificial Neural Network Predictor

All the neural network architectures used in this paper are feed-forward networks that learn by adjusting their weights according to the back-propagation algorithm as described in [14]. The number of neurons in the input layer is set equal to the number of bits in the binary inputs strings for the data set used. Each neuron of the input layer is set to either 0 or 1 depending on the value in the input string that it corresponds to. The values then propagate forward through the network until a pattern appears on the neurons of the output layer. The differences between them are propagated back through the network and the connection weights between the layers are adjusted. The threshold function used in the neurons is the unipolar sigmoid function given by,

$$f(x) = 1/1+\exp(-x)$$

The training patterns are input to the network in random order, to prevent the network from being skewed towards the last pattern presented. After each pattern has been input to the network, the difference between the network and desired output is used to update the weights on the network connections. The change in weights is given by

$$\Delta Wij\ k+1 = c\ (di-oi)\ oi\ (1-0i) + m\Delta wij\ k$$

where, Wij is the connection weight between neuron j on one layer with neuron I on the previous layer and di and oi are the desired and actual network output, respectively, for bit i. The constants c and m denote the learning rate and momentum of the network and they govern how fast the weights will change at each step and how much that change will depend on the previous weight update. Training continues until the sum of squares (S.S) error is less than some critical value (ecrit). The value of Sum of square is calculated from

$$S.S = \Sigma p \Sigma l\ (di-oi)\ 2$$

where, P is the total number of patterns presented. In the work presented here, 'ecrit' was chosen to have a value of 1.0. [14]

After training has been completed the learning is switched off and previously unseen test input patterns are presented to the network. These input patterns are propagated forwards to obtain an output. This output is the prediction of the following 24h minimum temperature category and, in the real application, would be used as an indicator as to whether frost prevention measures should be taken. In this investigation, the network output is compared to the output string associated with the input, to measure the network's performance.

Learning and Training

Learning and training are fundamental to nearly all neural networks. Training is the procedure by which the network learns; learning is the end result of that procedure. Learning consists of making systematic changes to the weights to improve the network's response performance to acceptable levels. The networks learn by adjusting the weights connecting the layers. The network starts by finding linear relationships between the inputs and the output. Weight values are assigned to the links between the input and output neurons. Once those relationships are found, neurons are added to the hidden layer so that nonlinear relationships can be found. The aim of training is to find a set of weights that will minimize error. During training, the output predicted by the network (Y (t)) is compared with the actual (desired) output (K (t)), and the mean squared error (MSE) between the two is calculated. The error function at time, Δ(t), is given by:

$$\Delta(t) = \tfrac{1}{2} \Sigma\ [Y\ (t)-K\ (t)]$$

The learning algorithm modified the weights associated with each PE such that the system minimizes the error between the target output and the network's actual output [15]. The back-propagation algorithm is the most computationally straightforward algorithm for training the MLP. More detailed explanation is available in most neural network text books [16]

ANN Applications in Aquaculture

Finfish and Shellfish Disease Forecasting Using ANN

A database was probed with artificial neural network (ANN) and multivariate logistic regression (MLR) models to investigate the efficacy of predicting PCR-identified human Adenovirus (ADV), Norwalk-like virus (NLV), and enter virus (EV) presence or absence in shellfish harvested from diverse countries in Europe (Spain, Sweden, Greece, and the United Kingdom). The relative importance of numerical and heuristic input variables to the ANN model for each country and for the combined data was analyzed with a newly defined relative strength effect, which illuminated the importance of bacteriophages as potential viral indicators [17]. The results of this analysis showed that ANN models predicted all types of viral presence and absence in shellfish with better precision than MLR models for a multicountry database. For overall presence/absence classification accuracy, ANN modelling had a performance rate of 95.9 per cent, 98.9 per cent, and 95.7 per cent versus 60.5 per cent, 75.0 per cent, and 64.6 per cent for the MLR for ADV, NLV, and EV, respectively. The selectivity (prediction of viral negatives) was greater than the sensitivity (prediction of viral positives) for both models and with all virus types, with the ANN model performing with greater sensitivity than the MLR. ANN models were able to illuminate site-specific relationships between microbial indicators chosen as model inputs and human virus presence. A validation study on ADV demonstrated that the MLR and ANN models differed in sensitivity and selectivity, with the ANN model correctly identifying ADV presence with greater precision.

Quality Control of Finfish and Shellfishes

When fish die, bacteria or the enzymes they produce invade the flesh of fish. This process produces toxic compounds in the fish and the fish becomes spoiled. Fourier Transform Infrared spectroscopy (FT-IR) allows chemical based non-destructive discrimination of intact microbial cells, and produces complex biochemical fingerprints which are reproducible and distinct for different accumulation of byproducts of spoilage. Data obtained through FT-IR spectroscopic analysis can be used to train an artificial neural network (ANN) for the development of an ANN based FT-IR Screening System for fish-quality inspection [18].

Eco-path Modelling of River Systems

Human activities have severely deteriorated the river systems, and many functions such as drinking water supply, fishing, etc. are threatened. Because the restoration of these river systems entails drastic social (*e.g.* change in habits with regard to water use and discharge, urban planning) and economical (*e.g.* investment in nature restoration, wastewater treatment system installation) consequences, the decisions should be taken with enough forethought. Ecosystem models could therefore act as interesting tools to support decision-making in river restoration management. In particular models that can predict the habitat requirements of organisms are of considerable importance to ensure that the planned actions have the desired effects on the aquatic ecosystems [19]. It was shown that machine learning techniques such as Artificial Neural Networks (ANNs) basically mimic aspects of biological information processing for data modelling and could be useful in ecology.

Predicts Antibiotic Residues in the Ecosystem

Fish antibiotics which are an important group of pharmaceuticals used in fish farming to treat infections and, until recently, most of them have been exposed to the environment with very little attention. Information about the environmental behaviour and the description of the environmental fate of medical substances are difficult or expensive to obtain [20].

Complex Systems of Regional Sustainable Development

Meeting the challenge of sustainable development requires substantial advances in understanding the interaction of natural and human systems. The dynamics of regional sustainable development could be addressed in the context of complex system thinking. Three features of complex systems are that they are uncertain, non-linear and self-organizing. Modelling regional development requires a consideration of these features. This paper discusses the feasibility of using the artificial neural network (ANN) to establish an adjustment prediction model for the complex systems of sustainable development (CSSD). Shanghai Municipality was selected as the research area to set up the model, from which reliable prediction data were produced in order to help regional development planning. A new approach, which could help to manage regional sustainable development, is then explored [21].

Ammonia Concentration Forecasting in an Eel (*Anguilla anguilla* L.) Intensive Rearing System

One of the main problems in the management of fish farms with water recirculating system is the forecasting and control of ammonia concentration in order to minimise the fish stress status. This paper examines methodologies of prediction in a real-time environment for an eel intensive rearing system. Approaches based on linear multiple regression, univariate time series models (exponential smoothing and autoregressive integrated moving average (ARIMA) models) and computational neural networks (ANNs) are developed to predict the daily average ammonia concentration in rearing tanks with water recirculating. The models are established using actual data from an eel fish farm in southern Spain. The input variables used in the models (multiple regressions, Holt smoothing model, ARIMA models and ANN models) are the ammonia concentration of previous days. In ANN models, the training method used is a standard back-propagation variation known as extended-delta-bar-delta (EDBD). Different neural architectures, whose learning is carried out by cross validation and controlling several threshold determination coefficients, are compared. Globally, the nonlinear ANN model approach is shown to provide a better prediction of daily average ammonia concentration than linear multiple regression and univariate time series analysis when the correlation between data series is low and when the models were obligated to predict in a situation for which specifically had not been calibrated [22]. The best results were obtained by 5:10s:15s:11 ANN model in the pre-growth series.

Fish Yield Map Modelling

Fish yield simulation is no longer restricted to predicting performance at a point. Models for pond dynamics and models for machinery, such as irrigation systems and harvesting equipment, allow us to predict the variability in fish production performance in the pond. Yield maps are now one of the most common sources of map data available to farmers for their fields. GPS units and ANN are a relatively low-cost addition to a new combine, and examples exist where farmers have used the yield/production maps to discover expensive production problems.

Satellite Image Processing

During 90's there have been considerable increases in both the availability of large remotely sensed data and the use of neural networks. This provides the opportunity to test the ability of neural networks, in particular the feed-forward back-propagation multi-layer perception, and compare the performance of particular neural networks with other traditional methods of satellite image processing The ANN approach has also been used to retrieve the correlation lengths and variance from rough surface to reconstruct the snow parameters to estimate the total culture area and also to estimate the production and productivity of water bodies.

Spatial Modelling with GIS

Recent research has shown that coupling ANN with GIS has significantly improved the modelling capabilities of GIS for spatial decision making In the pioneering work by the successfully strengthened the spatial data modelling capabilities of GIS for aquaculture land suitability analysis by integrating a neural network into a GIS environment.

Future Perspective and Values

In general, ANN technology mimics the brain's own reasoning process. The advantage of ANN for yield prediction has been motivated by the realization that the human brain is very efficient at processing large quantities of data from a variety of different sources, and making decisions in a complex environment. As humans apply knowledge gained from past experience to new problems or situations, a neural network takes previously solved examples to build a system of "neurons" that makes new decision, classifications, and predictions accurately and rapidly. In particular, the ANN approach shows advantages over statistical modelling approaches traditionally used to study natural systems [13]; Artificial Neural Network's

- ✫ it deal with the non-linearity's in the world in which we live
- ✫ It handles noisy or missing data.
- ✫ It creates their own relationship amongst information and no equation. They do not require *a priori* knowledge of the underlying process or assumptions of the structure of the target function.
- ✫ It can work faster than other when the problem is extremely complex or large numbers of variables or parameters. The neural network can develop its own weighting scheme based on relationships between the variables,

thus reducing the requirement that user provide all known information about a problem.

- ☆ it provide general solutions with good predictive more accurate than other statistical techniques, particularly when the problem or task addressed is either poorly defined or misunderstood, and observations of the process may be difficult or impossible to perform using incomplete data;
- ☆ Once trained, the nets can be used to analyze new conditions and provide suggested solutions. The ability of the net to learn complex relationships and the capability of including both qualitative as well as quantitative data makes the neural net approach a very flexible and powerful tool.

However, it is equally important to understand the basic problems of neural networks. There are three issues to be aware of, particular for those who are new to the use of neural networks. ANN is usually treated as a "black-box", with which the weights are uninterruptible due to presence of hidden layers and the nonlinearity of the activation function. Neural nets are not self-explanatory; there are no standard tests can measure the degree of variability in the outputs explained by certain inputs or the significance level of the predictions. This is one of reasons that forest managers are less likely to use ANN when a more familiar and better-understood procedure such as a regression analysis is available. On other hand, the research on ANN applications has been limited compared to other AI techniques (*e.g.*, Knowledge based System) in natural resource management [8]. There is an urgent need to widely recognize the potential uses of ANN as an alternative tool in fisheries field.

In this research an artificial neural network was employed to model the relationship between yield and the factors influencing yield. ANN techniques have been proven as a useful tool for predicting, classifying, and approximating functions in different fields, and are finding a wide range of applications in yield prediction and stock prediction. The utility of the ANN approach are apparent in applications (i) It is entirely numerical method as opposed to the implicit symbolism in the system of rules where the problem addressed may be either poorly understood, or observations of the process may be difficult to carry out using noisy or incomplete data; and (ii) It does not demands a previous knowledge of the system that we wish to predict and it is characterized when the problem is extremely complex, particularly when dealing with nonlinear. Systems, where traditional statistical techniques or mathematical models cannot to be formulated.

However, the well-known disadvantage of the neural network methods are i) It does not contributes knowledge about the background of the problem, while rules based system are more easily comprehensible. ii) In the majority of cases it does not permits an incremental training and including uninterruptible black box components; numerous training time, possible data over fitting and unsuited to such task (eg. Calculating Pay roll) we should balance its strengths against limitations when compared to traditional statistical techniques. The discipline of ANN not a panacea and still it is immature. It will not replace traditional quantitative techniques completely.

Fuzzy Logic Application in Fisheries and Aquaculture Research

Fuzzy logic is a problem-solving control system methodology that lends itself to implementation in systems ranging from simple, small, embedded micro-controllers to large, networked, multi-channel PC or workstation-based data acquisition and control systems. It can be implemented in hardware, software, or a combination of both. Fuzzy Logic provides a simple way to arrive at a definite conclusion based upon vague, ambiguous, imprecise, noisy, or missing input information. Fuzzy logic's approach to control problems mimics how a person would make decisions, only much faster.

Fuzzy logic is a superset of conventional (Boolean) logic that has been extended to handle the concept of partial truth— truth values between "completely true" and "completely false". It was introduced by Dr. Lotfi Zadeh of UC/Berkeley in the 1960's as a means to model the uncertainty of natural language. It is often suggested that the power of the human brain is a function of its ability to efficiently process imprecise information. To introduce this "fuzzy logic" to a computational setting, the critical elements are not numbers but fuzzy sets. Fuzzy sets allow for quick processing of information by association of vaguely similar patterns while providing the means to deal scientifically with subjectivity–a territory that traditional science has essentially ignored.

In other words, fuzzy logic is concerned with pulling away from logic that is crisp or Boolean (binary; 0 or 1). This method has an advantage over Boolean logic in that it mimics complex human reasoning in order to arrive at realistic conclusions about the imprecise and often fuzzy nature of reality. Openshaw suggests that fuzzy logic techniques provide the ability to develop soft computing applications that permit computer models to be specified and built from linguistic statements, based on common sense or theory or rules of thumb, it also offers a refreshingly new perspective on how to go about building better models of geographical systems by handling rather than ignoring or artificially removing the fuzziness within them (Openshaw and Openshaw 1997: 268-269).

To illustrate this concept, consider the thickness of a textbook. First of all, what is meant by the word thick? Using a crisp Boolean approach we may conclude, from a relative perspective, that anything over 4cm is thick. Consequently, a text with a thickness of 3.999cm is placed into the category of not thick. Clearly, this type of logic only works for a phenomenon that is non-continuous.

The fuzzy set approach considers a membership function of the thick set as a gradation between certainly not thick to certainly thick. Surely, a text that is 1cm thick would not be included in the set. However, as the number of pages in the text increases the degree of membership of the thick set becomes closer to 1. Therefore, instead of a vertical cut off line for thick membership at 4cm, a linear function depicts how a fuzzy membership might be approximated.

How is Fuzzy Logic Different from Conventional Control Methods

Fuzzy logic incorporates a simple, rule-based IF X AND Y THEN Z approach to a solving control problem rather than attempting to model a system mathematically.

The FL model is empirically based, relying on an operator's experience rather than their technical understanding of the system. For example, rather than dealing with temperature control in terms such as "SP =500F", "T <1000F", or "210C <TEMP <220C", terms like "IF (process is too cool) AND (process is getting colder) THEN (add heat to the process)" or "IF (process is too hot) AND (process is heating rapidly) THEN (cool the process quickly)" are used. These terms are imprecise and yet very descriptive of what must actually happen. Consider what you do in the shower if the temperature is too cold: you will make the water comfortable very quickly with little trouble. Fuzzy logic is capable of mimicking this type of behavior but at very high rate.

How Does Fuzzy Logic Work

Fuzzy logic requires some numerical parameters in order to operate such as what is considered significant error and significant rate-of-change-of-error, but exact values of these numbers are usually not critical unless very responsive performance is required in which case empirical tuning would determine them. For example, a simple temperature control system could use a single temperature feedback sensor whose data is subtracted from the command signal to compute "error" and then time-differentiated to yield the error slope or rate-of-change-of-error, hereafter called "error-dot". Error might have units of degs F and a small error considered to be 2F while a large error is 5F. The "error-dot" might then have units of degs/min with a small error-dot being 5F/min and a large one being 15F/min. These values don't have to be symmetrical and can be "tweaked" once the system is operating in order to optimize performance. Generally, Fuzzy logic is so forgiving that the system will probably work the first time without any tweaking.

Applications in Fisheries

Water Pollution Regulations

Closed-system aquaculture presents a new and expanding commercial opportunity; however, presently there are few profitable commercial operations. Intensive aquaculture systems require a sufficient supply of oxygen and nutrients in temperature- controlled water with a means of removing wastes from the system.In properly controlled recirculating aquaculture systems, stocking rates of optimum quantities of fish per gallon of water can be obtained. However, since failure of any component can cause catastrophic losses within a short period of time, the system must be reliable and constantly monitored. Thus, precise measurements and controls are necessary for the success of an intensive recirculating aquaculture system. An important component of closed-system aquaculture is the control system, which must measure and control all of the critical system parameters. Recent developments in control technology and microcomputers may revolutionize the operation and control of many agricultural and biological production systems. A properly controlled system will also be energy efficient since production can be optimized with respect to the various inputs.

The most important parameters to be monitored and controlled in an aquaculture system are related to water quality, since they directly affect animal health, feed utilization, growth rates and carrying capacities. The primary water quality parameters include temperature, dissolved oxygen (DO), pH, ammonia, nitrites, nitrates, suspended solids, salinity, alkalinity, biochemical oxygen demand (BOD), and water flow rate. Water levels, availability of electric power, means of detecting fire, smoke and the intrusion of vandals should also be monitored.

One of the major problems facing modern filtration technology in aquaculture is the cost-effective filtration of nitrate, NO_3. The design of a cost-effective, integrated nitrogen removal system including both nitrification and denitrification processes is paramount to the continued development of commercial recirculating aquaculture systems. Current nitrification systems are adequate and many are commercially available; however, current techniques for denitrification are almost exclusively based on water exchange. A fuzzy logic system to control the operation of a denitrification bioreactor has been developed by the University of Texas Medical Branch, Marine Biomedical institute; USA for use in a closed recirculating aquaculture system. The development of a prototype system requires a multi-disciplinary program involving computer scientists, bioengineers, biochemists and aquaculturists.

Detecting Environmental Regimes in Fish Stock-Recruitment Relationships

Fuzzy logic approach is developed to model and test the impact of environmental regimes on fish stock–recruitment relationships. Traditional methods use environmental variables to classify stock–recruitment data into different membership percentiles followed by fitting the stock–recruitment models for each subset. In contrast, the fuzzy logic approach uses a continuous membership function to provide a rational basis for the classification. Thus, parameter estimation is based on a more logically consistent foundation without resorting to subjective partitions. This new approach is applied to herring stock from the west coast of Vancouver Island (*Clupea harengus pallasi*) using sea surface temperature as the environmental variable and to Pacific halibut stock (*Hippoglossus stenolepis*) using the Pacific Decadal Oscillation as the environmental variable (D.G.Chen, 2001). From these applications, the herring stock–recruitment relationships were found to vary significantly during different regimes, whereas this was not the case for halibut. However, in both instances, the fuzzy logic approach demonstrated that density-dependent effects differed between regimes. The fuzzy logic model consistently outperformed traditional approaches as measured by several diagnostic criteria. Because fuzzy logic models address uncertainty better than traditional approaches, they have the potential to improve our ability to understand factors influencing stock–recruitment relationships and thereby manage fisheries more effectively.

Fish Stock Assessment and Management in Open Seas

A new fuzzy logic model with a genetic algorithm is developed that overcomes some of the inherent uncertainties in the fish stock-recruitment process in open water bodies. This model is applied to stock-recruitment relationships for the Southeast Alaska pink salmon (*Oncorhynchus gorbuscha*) and the West Coast Vancouver Island Pacific herring (*Clupea pallasi*) stocks (D.G. Chen *et al.*, 2000). In both examples, the annual mean sea surface temperature is used as an environmental intervention in the model. The fuzzy logic model provides the functional relationship between the number of fish spawners and the sea surface temperature that is used to reconstruct the historical fish recruitment time series and also to predict the number of fish that will recruit in the future. Globally optimized genetic learning algorithms are used to find the optimal values of the parameters for the fuzzy logic model. The results from this fuzzy logic model are compared with results from both a traditional Ricker stock-recruitment model and a recent artificial neural network model. These comparisons demonstrate the superior capability of the fuzzy logic model for addressing problems of uncertainty and vagueness in both the data and the stock-recruitment relationship. The fuzzy logic model approach is recommended as a useful addition to the analytical tools currently available for fish stock assessment and management.

Conclusion

This chapter focuses on the recent trends in ANN and Fuzzy logic applications in Aquaculture. It is observed that the fuzzy logic model consistently outperformed traditional approaches as measured by several diagnostic criteria. Because fuzzy logic models address uncertainty better than traditional approaches, they have the potential to improve our ability to understand factors influencing stock–recruitment relationships and thereby manage fisheries more effectively. A complete interpretation of the fuzzy scores will not be possible until additional data and environmental information can be collated and analyzed. From the above discussion it is clear that plausible relationships exist between the fuzzy memberships calculated from the visual data in the logs, and environmental factors that are likely to affect benthic conditions. On the basis of this preliminary work we feel that the use of fuzzy logic to analyse these data appears to be successful.

References

1. Lo, J. Y.; Baker, J. A.; Kornguth, P. J.; Floyd, C. E. 1995. Application of artificial neural networks to interpretation of mammograms on the basis of the radiologists impression and optimized image features. *Radiology* 197: 2A2-2M.

2. Colasanti, R. L. 1991. Discussions of the possible use of neural network algorithms in ecological modelling. *Binary* 3: 13-15.

3. Edwards, M.; Morse, D. R. 1995. The potential for computer-aided identification in biodiversity research. *Trends in Ecology and Evolution* 10: 153-158.

4. Scardi, M. 1996. Artificial neural networks as empirical models for estimating phytoplankton production. *Marine Ecology Progress* Series 139: 289-299

5. Brosse, S.; Guegan, J. F.; Tourenq, J. N.; Lek, S. 1999b. The use of artificial neural networks to assess fish abundance and spatial occupancy in the littoral zone of a mesotrophic lake. *Ecological Modelling,* 120: 299-312.

6. Brosse, S.; Lek, S.; Dauba, F. 1999c. Predicting fish distribution in a mesotrophic lake by hydroacoustic survey and artificial neural networks. *Limnology and Oceanography* 44: 1293-1303.

7. Schleiter, 1. M.; Borchardt, D.; Wagner, R.; Dapper, T; Schmidt, K. D.; Schmidt, H. H.; Werner H. 1999. Modelling water quality, bioindication and population dynamics in lotic ecosystems using neural networks. *Ecological Modelling* 120: 271-286

8. Coulson, R. N., L. J. Folse and D. K. Loh, 1987. Artificial intelligence and natural resource management. *Science* 237: 262-267.

9. Pham, D. T., X. Liu, 1995. *Neural Networks for Identification: Prediction and Control*. Springer-Verlag London Limited, Great Britain. 238 p.

10. Brown, D., P. Rothery, 1993. *Models in Biology: Mathematics, Statistics and Computing*. John Wiley and Sons Ltd., England. 688 p.

11 George, R. K., S. Rammohan, M. S. Kulshretha, A. M. Shekh and H. Jaita, 2001. Prediction of soil temperature using neural networks. *Journal of Agrometeorology* 3(1&2):169-173. India

12. McCulloch, W.C and W. Pitts, 1943. A logical calculus of the ideas immanent in nervous activity, *Bulletin of Mathematical Biophysics* 5:115-133.

13. Atkinson, P. M. and A. R. Tatnall, 1997. Introduction: Neural networks in remote sensing, *Int. J. Remote Sensing* 18:699-709.

14. Zurada, J. M., 1998. *Introduction Artificial Neural Systems*. Jaico publishing house, New Delhi, India.

15. Changhui Peng1 and Xuezhi Wen2, 1999. Recent Applications of Artificial Neural Networks in Forest Resource Management: An Overview *In*: *Environmental Decision Support Systems and Artificial Intelligence*, Ulises Corté and Miquel Sànchez-Marrè, Cochairs (eds.).pp.15-22.

16. Technical Report.1999.WS-99-07, AAAI Press, Menlo Park, CA.

17. Gail Brion, Chandramouli Viswanathan, T. R. Neelakantan, Srinivasa Lingireddy, Rosina Girones, David Lees, Annika Allard and Apostolos Vantarakis, 2005. Artificial Neural Network Prediction of Viruses in Shellfish. *Applied and Environmental Microbiology*, Sept. 2005, 71(9):5244–5253.

18. Singhal, J. Zhang, H. E. Michel and B. R. Singh,.2002. Fish-Quality Analysis using Artificial Neural Networks and Spectroscopic Data. *Proceedings of International Conference on Artificial Intelligence and Soft Computing*, ASC Banff, Canada P. 357-402.

19. Andy P., Dedecker, Peter L.M. Goethals, Wim Gabriels, Niels De Pauw, 2004. Optimization of Artificial Neural Network (ANN) model design for prediction of macro invertebrates in the Zwalm river basin (Flanders, Belgium). *Ecological Modelling* 174:161–173.

20. Maran, E., M., NoviÄ,ˋ P. Barbieri, J. Zupan, 2004. Application of counter propagation artificial neural network for modelling properties of fish antibiotics. *SAR and QSAR in Environmental Research*, 15(5-6):469-480.

21. Shi Chun, Philip James and Guo Zhong-yang. 2004. Application of artificial neural network in complex systems of regional sustainable development. *Chinese*

Geographical Science, Science Press, co-published with Springer-Verlag GmbH, 14(1/March): 1-8.

22. Juan C. Gutiérrez-Estrada, Emiliano de Pedro-Sanz, Rafael López-Luque and Inmaculada Pulido-Calvo. 2004. Comparison between traditional methods and artificial neural networks for ammonia concentration forecasting in an eel (Anguilla anguilla L.) intensive rearing system. *Aquacultural Engineering*, 31(3-4): 183-201.
23. D.G. Chen, N.B. Hargreaves, D.M. Ware, and Y. Liu. 2000. Can. J. Fish. Aquat. Sci./J. can. sci. halieut. aquat. 57(9): 1878-1887.
24. D.G.Chen. 2001. Can. J. Fish. Aquat. Sci./J. can. sci. halieut. aquat. 58(11): 2139-2148.
25. Europe Gets into Fuzzy Logic" (*Electronics Engineering Times*, Nov. 11,1991)
26. L.A. Zadeh. 1987. (Ed.) R.R. Yager *et al.* (John Wiley) *Fuzzy Sets and Applications:* Selected Papers, New York.
27. Openshaw, Stan and Christine Openshaw. 1997. *Artificial Intelligence in Geography*. John Wiley, Chichester.

Modelling, Forecasting, Artificial Neural Network and Expert System in Fisheries
Editors: Ajit Kumar Roy & Niranjan Sarangi
Published by: DAYA PUBLISHING HOUSE (2009)

Pages 206–219

Chapter 15

Soft Computing Techniques for Data Analysis and Modelling in Fisheries

H.C. Verma

Introduction

The soft computing is a collection of computational techniques based on the information processing in biological systems. The role model for soft computing is the human brain. Human type information processing involves both logical and intuitive information processing. Conventional (hard) computing is good for formal logical information processing and relies heavily on computer-aided numerical analysis, but their capability for the intuitive information processing is far behind that of human beings. Soft computing differs from hard (conventional) computing in that, unlike hard computing, it is tolerant of imprecision, uncertainty, partial truth, and approximation. Soft computing techniques are flexible enough to support three features: openness, robustness, and real time processing. Openness of a system is its ability to adapt or extend on its own to cope with changes encountered in the real world. Robustness of a system means its stability and tolerability when confronted with distorted, incomplete, or imprecise information. The real time characteristic implies the ability of the system to react within a reasonable time in response to an event. The basic ideas underlying soft computing in its current form have links to the work by Zadeh (1965) on fuzzy sets. Key areas of soft computing include: Artificial Neural Networks (ANN), Support Vector Machines (SVM), Evolutionary Algorithms (EAs), Fuzzy Sets (FS), Wavelets, Rough Sets (RS), Simulated Annealing (SA), Swarm intelligence, Bayesian network and Chaos theory. These techniques attempt to study, model, and analyze very complex phenomena for which more conventional methods have not yielded low cost, analytic, and complete solutions. Artificial neural networks and support vector machines are also called machine-learning techniques. In recent years, machine learning has become a focal point in Artificial Intelligence. Recently, use of soft computing techniques for solving bioinformatics problems have been gaining the attention of researchers because of their ability to handle imprecision, uncertainty and near optimality in large and complex search spaces.

In this chapter, the basic principles of some soft computing techniques *viz.*, neural networks and support vector machines, have been described and their implementation details have been discussed with examples.

Artificial Neural Networks

The development of artificial neural networks began approximately 60 years ago (McCulloch and Pitts 1943), inspired by a desire to understand the human brain and emulate its functioning. A significant growth in the interest of this computational mechanism has occurred since Rumelhart *et al.* (1986) developed a mathematically rigorous theoretical framework for neural networks. Artificial Neural network (ANN) is an information processing paradigm inspired by biological nervous systems, such as our brain. It simulates human neural system. It consists of large number of highly interconnected processing elements (called neurons) working together (Tsoukalas and Uhrig 1997). Once trained, the neural network is able to recognize similarities when presented with a new input pattern, resulting in a predicted output pattern. Neural Networks are very efficient in identifying quite complex non-linear relationships between the input and output (Rao and Rao 1996). They have ability to deal with incomplete and noisy data. It learns from past experiences then generalizes to unseen patterns. ANN is considered as one of the most active areas of artificial intelligence. With the growing computing power, ANN is becoming a powerful tool for forecasting and data analysis. Artificial neural networks are very helpful when we can't formulate an algorithmic solution; we can get lots of examples of the behavior we require; and we need to pick out the structure from existing data.

Architecture and Mathematical Representation

ANN consists of a large number of processing elements, called neurons. Each neuron has an internal state, called its activation or activity level which is a function of the inputs it has received. Typically, a neuron sends its activation as a signal to several other neurons. The network consists of several "layers" of neurons: an input layer, hidden layer(s), and an output layer (Figure 15.1). Input layer takes the input and distribute it to the hidden layers. These hidden layers do all the necessary computation and output the results to the output layer, which outputs the data to the user. A neural network (NN) is characterized by architecture–its pattern of connections between the neurons; learning algorithm–its method of determining the weights on the connections; and activation function–which determines its output (Fausett 1994).

Consider $x_1, x_2, x_3.....x_n$ are the n inputs to the neuron 'j' and forms an input vector $X=(x_1, x_2, x_3.....x_n)$ which is also called input pattern. The sequence of weights leading to the node 'j' forms a vector $W_j = (w_{1j}, w_{2j}, w_{3j}, ...w_{nj})$, where w_{ij} represents the connection

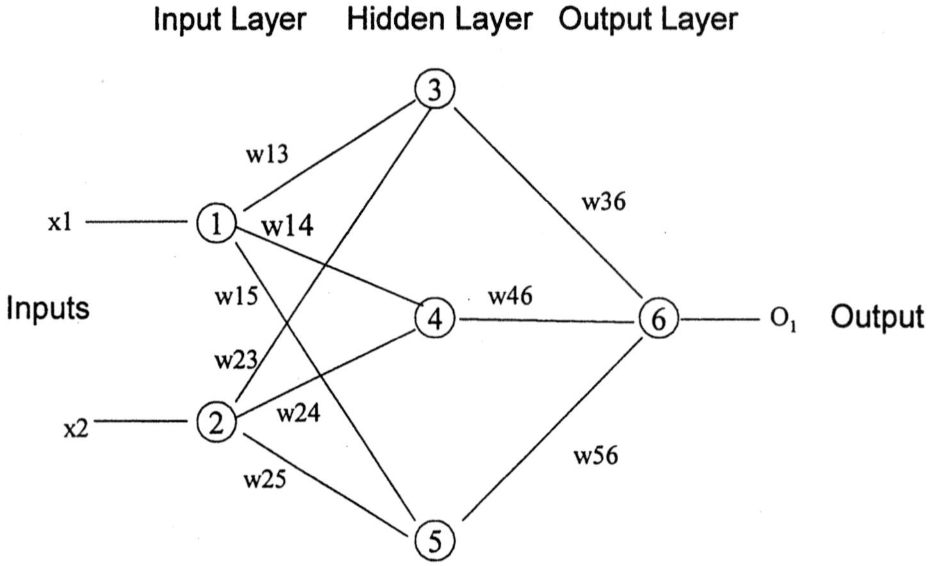

Figure 15.1: Configuration of Feed Forward Three-Layer ANN

weight from the ith node in preceding layer to this node. Here the total input I received by node 'j' is–

$$I = \sum_{i=1}^{n} w_{ij} x_i \tag{1}$$

In Figure 15.1, the value passed from neuron 3 to neuron 6 is = f(w13* value of neuron 1+ w23* value of neuron 2).

To generate the final output O_j, the sum is passed on to non-linear filter f(.) called Activation/Transfer/Squash function.

$$O_j = f(I-\theta) \tag{2}$$

where, θ is threshold value. The transfer functions do the final mapping of the activations of the output neurons into the network output. The neuron calculates a weighted sum of inputs and compares it to a threshold. The output is computed based on the type of transfer function. There are different forms of f(0.) used in eq. (2) as shown in Figure 15.2.

The output from a single cycle of operation of a neural network may not be the final output, since the network would be iterated into further cycles of operation until the network output converges. In order for an ANN to generate an output vector $O=(o_1,o_2,o_3....o_n)$ which is as close as possible to the target output vector, say $T=(t_1,t_2,t_3,.....t_n)$, a training process must be employed to find optimum weight matrix W, that minimizes a predetermined error function E given by

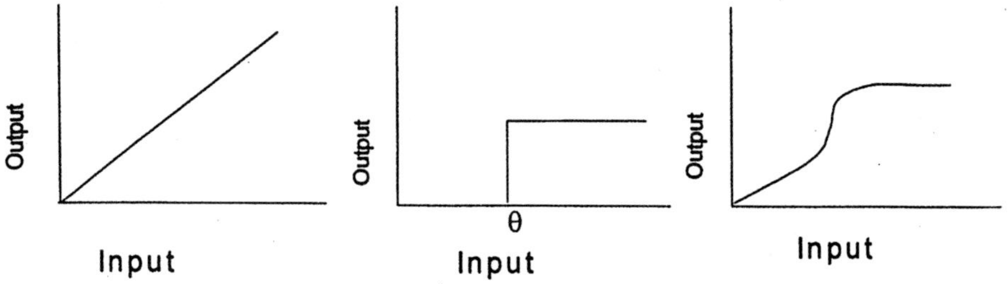

Linear function: $f(x)=wx_j+b$

Step function: $f(x)=\begin{cases}0 & \text{if } x<\theta \\ 1 & \text{if } x>=\theta\end{cases}$

Sigmoid function: $f(x)=1/(1+e^{-x})$

Figure 15.2: Activated Functions

$$E=\sum_{p}\sum_{n}(o_t-t_i)^2 \qquad (3)$$

where, n= no. of output nodes and p= no. of training (input/output) patterns.

ANN Models

Over the years, several neural network systems have evolved. The following are some of the systems that have gained significance:

- ☆ Perceptron
- ☆ Multi-Layer Feed Forward Network (MLFF)
- ☆ Radial Basis Function (RBF)
- ☆ Time Delay Neural Network (TDNN)
- ☆ Counter Propagation Network (CPN)
- ☆ Recurrent Neural Network (RNN)
- ☆ Hopefield Network
- ☆ Adoptive Resonance Theory (ART)
- ☆ Fuzzy Associative Memory (FAM)
- ☆ Bidirectional Associative Memory (BAM)
- ☆ Self Organizing Feature Maps (SOFM)

Learning and Training in Neural Network

The aim of learning is to map a set of inputs to a desired set of output, either with or without an explicit training process. Neural networks "learn" from experience/ examples. The network can also learn in the absence of training. In a biological system, learning involves adjustments to the synaptic connections between neurons. The same is true for artificial neural networks. The strength of connection between the neurons is stored as a weight value for the specific connection. The system learns new knowledge by adjusting these connection weights. The learning ability of a neural network is determined by its architecture and by the algorithmic method chosen for training. Training is an external process but learning is desired process that takes place internal to the network (Rao and Rao 1996).

Learning Methods and Rules

Learning methods in neural networks can be broadly classified into three basic types: unsupervised, supervised, and reinforcement (Rajsekharan and Pai 2003). In unsupervised learning, the hidden neurons must find a way to organize them without help from the outside. This is often referred to as self-organization or adoption. In this approach no training data/information is made available on the desired output. It is learning by doing. It is used to pick out structure in the input such as clustering and classification. It may also use competitive learning where the weights of only winning neuron are updated. The models that use unsupervised learning are Kohonen Self-organizing Map, Hopefield Network, Bidirectional Associative Memory, Fuzzy Associative Memory and ART. In supervised learning method, both the inputs and the outputs are provided. The network then processes the inputs and compares its resulting outputs against the desired outputs. Errors are then calculated, causing the system to adjust the weights, which control the network. This process occurs over and over as the weights are continually tweaked.The ANN models that use supervised learning are Feed Forward Back Propagation Network, Adaline, Radial Basis Function networks and Perceptions. In reinforcement learning method, the teacher though available, doesn't present expected answer but only act as an observer who grades the performance of the network results.

Supervised and unsupervised learning methods, which are most popular forms of learning, have found expression through various rules. Hebb's rule was proposed by Hebb (1949) is used for unsupervised learning. It is based on corrective weight adjustment and can be expressed as

$$\Delta w_{ij}=\eta a_i a_j$$

where, η is learning rate (a changeable value used by several learning algorithms, which effects the changing of weight values. It has positive value much smaller than 1), w_{ij} is the connection weight between neuron i and j, a_i is the activation of ith neuron and a_j is the activation of jth neuron to the connection weight between i^{th} and j^{th} neuron.

Gradient descent rule is supervised training algorithm. It is based on minimization of error E defined in terms of weights and the activation function of the network.

$$\Delta w_{ij}=\eta . \partial E/\partial w_{ij}$$

where, Δw_{ij} is weight update of the link connecting the ith and jth neurons of two neighbouring layers; $\partial E/\partial w_{ij}$ is error gradient with reference to the weight w_{ij}.

Feed Forward Back Propagation Network

Back propagation is a supervised learning algorithm and is mainly used by Multi-Layer-Perceptions to change the weights connected to the net's hidden neuron layer(s). The back propagation algorithm uses a computed output error to change the weight values in backward direction. To get this net error, a forward propagation phase must have been done before. It is called Feed forward Back propagation Network (Figure 15.1). While propagating in forward direction, the neurons are being activated using the sigmoid activation function. The first input pattern is propagated through the net. The same procedure is used for the next input pattern, but then with the changed weight values. After the forward and backward propagation of the second pattern, one learning step is complete and the net error can be calculated by adding up the squared output errors of each pattern. By performing this procedure repeatedly, this error value gets smaller and smaller. The algorithm is successfully finished, if the net error is zero or approximately zero.

Self-organizing Feature Maps (SOFM) Network

SOFM, originally proposed by Kohonen (1990) has found widespread use as a method of mapping for high-dimensional data to low dimension. It uses unsupervised learning algorithm called self-organization. A neural net tries to simulate the biological human brain, and self-organization is probably the best way to realize this. The neural cells organize themselves in groups, according to incoming information. Those incoming information are not only received by a single neural cell, but also influences other cells in its neighborhood. This organization results in some kind of a map, where neural cells with similar functions are arranged close together. A neural network can also perform this self-organization process. These neural nets are mostly used for data clustering purposes, because similar input values are represented in certain areas of the net's map.

Keys Steps of Applying ANNs

Data Preparation

Input data selection and preparation are the most time-consuming tasks of neural network. A neural network is only as good as the input data used to train it. Once the most appropriate raw input data has been selected, it must be prepossessed. Transformation and normalization are two widely used preprocessing methods. Transformation involves manipulating raw input data to reduce its dimensionality (reducing number of input variables), while normalization is a transformation performed on input data to distribute the data evenly and scale it (mostly in the range of –1 to 1 or 0 to 1) into an acceptable range for the network. Some simple preprocessing methods include computing differences between or taking ratio of inputs. This reduces number of inputs to the network and helps it learn more easily.

Networks Selection

It involves selection of network model, number and size of hidden layers, initial weight matrix etc.

Training the Network

Much of the process of determining the best parameters for a given application is trail and error. It needs to start with a network topology and initial weight, and train the network on your training data set. When the network reaches the satisfactory minimum error, save the weights.

Testing and Interpretation of Results

The trained network is applied on test data set to find the error on this data set. If it is not satisfactory the training process needs to be repeated again with different set of initial weights. If the error is not satisfactory then you need to re-enter the design phase or the train and test phase.

Applications of ANNs

Neural networks are a powerful technique to solve many real world problems. They have the ability to learn from experience in order to improve their performance and to adapt themselves to changes in the environment. There are many applications of ANNs in the field of fisheries such as fish yield prediction (Lae *et al.*, 1999), parametrizing long term fisheries data (Hyun *et al.*, 2005), and prediction of fish food consumption (Ruohonen 1999; Lek 1995) etc. Many nets are used to implement expert systems, such as to give medical diagnoses and treatment. ANNs have become popular as pattern recognition procedures in sequence analysis (Wang *et al.*, 1998). Neural networks have been demonstrated to be convincingly superior to simple cost-matrix algorithms or Hidden Markov Models of sequence data. Nonetheless they are somewhat more flexible in implementation and interesting procedures exist for self-training neural nets that may even be useful in data reduction or machine learning, by extracting relevant information. Neural networks have been successfully applied for the solution of a variety of problems.

However, some of the common application domains are prediction such as prediction of viruses in Shellfish, prediction of fish yield, fish quality analysis, prediction of protein secondary structure, environmental risk forecasting etc.; classification and pattern recognition such as discovery of gene expressions, classification of molecular sequences, image processing, pattern matching, medical diagnosis etc.; data association and Data filtering. Some examples from the field of fisheries, where ANN has been successfully applied, are discussed below in detail:

Shrimp Disease Occurrence Prediction

Predicting the occurrence of disease outbreaks in aquacultural farms can be of considerable value to the long-term sustainable development of the industry. Prior research on disease prediction has essentially depended upon traditional statistical models with varying degrees of prediction accuracy. Leung and Tran (2000) have developed a neural network model to predict shrimp disease outbreaks in Vietnam using farm-level data from 480 Vietnamese shrimp farms. The ANN model was developed with 68 inputs that includes information about age of farm, nature of aquaculture activities, type of land use, soil type, operation, water source, shrimp species, farming techniques, water, sediment, diseases and their consequences, inputs, costs, revenue, production and profit trends. They evaluated the potential of ANN as an alternative to the traditional logistic regression model for the purpose of predicting disease occurrence in shrimp farms and the results indicated that ANN can outperform the logistic regression model when the full set of input variables was used.

Predicting Fish Yield

Lae *et al.* (1999) applied artificial neural network (ANN) to modelling and prediction of fish yield, using data from 59 lakes that are distributed all over Africa and Madagascar, as related to the environmental characteristics from the combination of six variables: catchment area over maximum area, fishing effort, conductivity, depth, altitude and latitude. The output variable was annual fish yield. Fish yields estimated with this method were significantly related to the observed fish yields with the correlation coefficient reaching 0.83.

Modelling Fish Food Consumption

Ruehonen (1999) has applied ANN for modelling fish food consumption. The model was developed using data relating to the daily consumption of commercial high-energy dry pellets. Data were collected for 161 days for the training set and for 119 days for the test set. He concluded that simple feed forward ANN may have potential for modelling the food consumption time series of fish, and that ANN could be used for site-specific feeding optimization based on experience from previous years.

Strengths and Limitations of Neural Networks

The following are some of the reasons ANNs have become attractive computational tools:

- ☆ Adapt to unknown situations.
- ☆ Model complex functions and have capabilities of universal function approximation and generalization.
- ☆ Deal with non-linearity of underlying process.
- ☆ Handle noisy data *i.e.* they work well even when the training sets contain noise and measurement errors.
- ☆ Work with large numbers of variables or parameters.
- ☆ Although several studies indicate that ANN has proven to be potentially useful tool to solve many complex problems, their limitation should not be ignored. Some of major limitations are:
- ☆ Large complexity of the network structure and chance of over fitting data.
- ☆ They work better when the data set is sufficiently large, therefore it is a data-hungry procedure.
- ☆ Neural networks require an extensive amount of training time.

Support Vector Machines (SVMs)

Support Vector Machines (SVMs) are a relatively new, general formulation for learning machines. Support vector machines (SVMs) are a set of related supervised learning methods used for classification and regression. Vapnik (1995) argued that to control the generalization ability of a learning machine one has to control the two different factors: the error rate on the training data and the capacity of the learning machine as measured by its Vapnik-Chervonenkis (VC) dimension which is a non-negative integer, measures the expressive power of the family of classification functions realized by the learning machine (Haykin, 2003). A special property of SVMs is that they simultaneously minimize the empirical classification error and maximize the geometric margin.Hence they are also known as maximum margin classifiers. A classification task usually involves with training and testing data, which consist of some data instances. Each instance in the training set contains one 'target value' called class labels, and several 'attributes' called features. The goal of SVM is to produce the model that predicts target value of data instances in the testing set, which are given only the attributes.

Mathematical Representation

Consider a finite training sample of N patterns $\{(x_i,y_i), i=1......N\}$, where x_i denote i^{th} pattern in n-dimensionl space *i.e.* $x_i=[x_{1i},x_{2i},........,x_{ni}] \in \Re^n$ and $y_i \in \{1,-1\}$ represents the class label of that pattern (in case of regression, $y_i \in \Re$ is the corresponding value of desired output). A non-linear transformation function $\Phi(.)$ is defined to map the input space to a higher dimensional feature space (Figure 15.3).

$$Y=f(x)=w^T\Phi(x)+b \qquad (4)$$

where, Y denotes the actual model output and the coefficients w and b are the adjustable model parameters ($w \in \Re^n$ and $b \in \Re$).

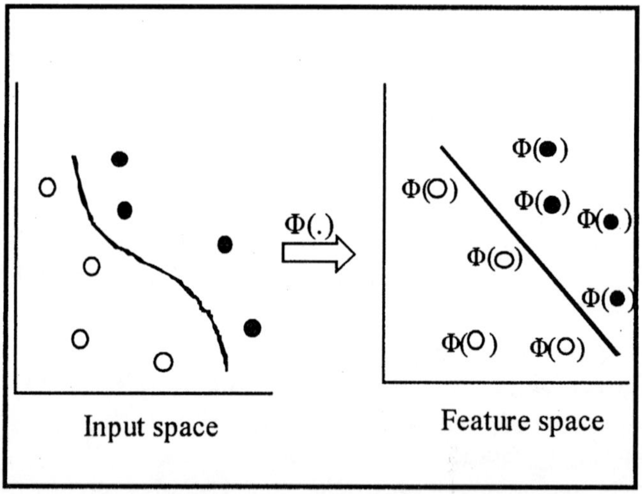

Figure 15.3: Mapping of Input Space to a Higher Dimensional Feature Space

Support vector machines map input vectors to a higher dimensional space where a maximal separating hyperplane is constructed. Two parallel hyperplanes are constructed on each side of the hyperplane that separates the data (Figure 15.4). In Figure 15.4, four open circles of class A and four filled circles of the class B are distributed in balance. Suppose a shaded circle belonging to class B is included, the hyper-plane (the dashed line) will be biased because the error (distance) between all thirteen circles and the hyperplane has to be minimised. In searching for the best

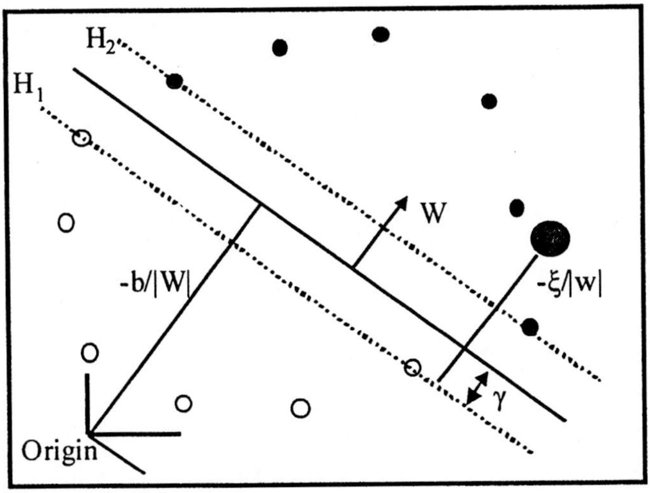

Figure 15.4: Linear Separating Hyperplanes for Non-separable Case

hyper-plane, SVMs find a set of data points that are the most difficult training points to classify. These data points are referred to as support vectors. In constructing SVM classifier, the support vectors are closest to the hyper-plane and are located on the boundaries of the margin between two classes. The advantage of using SVMs is that the hyper-plane is searched through maximising this margin. Because of this, the SVM classifier is the most robust, and hence has the best generalisation ability. In Figure 15.4, two open circles on the upper boundary and two filled circles on the lower boundary are selected as support vectors. Only the use of these four circles can form the boundaries of the maximum margin between two classes. Suppose we have some hyperplane, which separates the positive from the negative examples. The points x which lie on the hyperplane satisfy $w.\ x + b = 0$, where w is normal to the hyperplane, b is offset parameter that allows to increase the margin, $|b|/||w||$ is the perpendicular distance from the hyperplane to the origin, and $||w||$ is the Euclidean norm of w. Let γ (gamma) be the shortest distance from the separating hyperplane to the closest positive (negative) example. The margin of a separating hyperplane is to be 2γ. For the linearly separable case, the support vector algorithm simply looks for the separating hyperplane with largest margin. This can be formulated as follows:

Suppose that all the training data satisfy the following constraints:

$$x_i.\ w + b >= 1 \text{ for } y_i = 1 \quad (5)$$

$$x_i.\ w + b <= -1 \text{ for } y_i = -1 \quad (6)$$

These can be combined into one set of inequalities:

$$y_i(x_i.\ w + b) - 1 >= 0 \text{ for every } i \quad (7)$$

Now consider the points for which the equality in Eq. (5) holds. These points lie on the hyper plane $H1{:}x_i.w+b=1$ with normal w and perpendicular distance from the origin $|1-b| = ||w||$. Similarly, the points for which the equality in Eq. (6) holds lie on the hyper-plane H2:x_i.w+b=-1, with normal w, and perpendicular distance from the origin $|-1-b| = ||w||$. Hence $\gamma=1/||w||$ and the margin is simply $2/||w||$. Thus we can find the pair of hyperplanes which gives the maximum margin by minimizing $||w||^2$, subject to constraints (7).

If there exists no hyperplane that can split the data linearly, the 'soft margin' method is used to choose a hyperplane that splits the data as clearly as possible, while still maximizing the distance to the nearest clearly split examples. The method introduces slack variables, ξ_i, which measure the degree of misclassification of the datum x_i

$$y_i(x_i \cdot w + b) \geq 1-\xi_i \; 1 \leq i \leq N \tag{8}$$

The objective function is then increased by a function which penalises non-zero ξ_i, and the optimisation becomes a trade off between a large margin, and a small error penalty. If the penalty function is linear, then objective function can be given as

$$\min ||w||^2 + C\Sigma\xi_i \tag{9}$$

where, constant C is an additional constraint on the Lagrange multipliers. Non-linear penalty functions have been used, particularly to reduce the effect of outliers on the classifier.

Kernal Function

The original optimal hyperplane algorithm proposed by Vapnik (1963) was a linear classifier. However, Boser *et al.* (1992) suggested a way to create non-linear classifiers by applying the kernel trick to maximum-margin hyperplanes. The trained SVM classifier is a linear combination of the similarity between an input and the support vectors. The similarity between an input and the support vectors is quantified by a kernel function defined as: $K(x, \chi_i)$, where χ_i is the ith support vector. The decision is made using the following equation:

$$Y=f(x) = \Sigma\alpha_i y_i K(x, \chi_i)+b \tag{10}$$

where, y_i is the class label of the ith support vector and α_i is the positive parameter of the ith support vector determined by an SVM algorithm.

The most difficult part in SVMs is the design of a proper kernel function that corresponds to the selection of a proper number of hidden neurons in neural networks. There have been many kernel functions designed for dealing with numerical attributes.

Some common kernels include:

- ☆ Linear: $K(x_i, x_j) = x_i^T x_j$
- ☆ Polynomial: $K(x_i, x_j) = (\gamma x_i^T x_j + r)^d, \gamma > 0$
- ☆ Radial Basis Function: $K(x_i, x_j) = \exp(-\gamma \| x_i - x_j \|^2), \gamma > 0$
- ☆ Sigmoid: $K(x_i, x_j) = \tanh(\gamma x_i^T x_j + r)$

Here γ, r and d are kernal parameters which have to be adjusted while training the SVM.

Applications of SVM

SVMs perform exceptionally well on function approximation and regression problems. SVMs have found wide application in many areas including pattern recognition, regression, bio-informatics and artificial intelligence. Biological applications of SVMs involve classifying objects as diverse as protein and DNA sequences, microarray expression profiles. Possible applications in bioinformatics and fisheries include: Fish age classification, grading fish from images, discriminant analysis of microarray data, and prediction of protein-protein interaction, protein structure. Some applications of SVM in field of fisheries are described below:

Fish Age Classification

Knowledge of age in fish populations is one of the most important issue in stock assessment and management. Age determination through the manual interpretation of growth patterns is complex, and extremely time-consuming since the accuracy depends on the researcher's skills. Although several million otolith macrostructures are manually examined every year worldwide, statistical learning techniques have been adopted to automate such analyses and classifications of otolith images in fish age classes. Sergio Bermejo (2007) applied support vector machine (SVM) for fish age classification based on length, weight, sex and otolith morphological features. One hundred and forty-five specimens of known ages (varying from 2 to 6 years of age) were employed. Four morphological features were taken from the silhouette image of each otolith: the contour length, the area, the maximum internal distance and the maximum perpendicular distance. Additionally, information was available about each sample's length, weight and sex. Sex was coded in a feature component as a discrete value, in which females were assigned a value of 1 and males a value of 0. They compared otholith morphological features versus other features like fish length, weight and sex. An extensive set of experiments was performed for the three multi-class SVM classifiers using the polynomial kernel and the Gaussian kernel, as well as various values for the soft margin of the SVM C parameter. The best results were obtained with RBF kernels and for 1-vs-1 SVM with $C=\infty$. The accuracy was tested for different support vector machine classifiers using the cod database. They found that fish length, weight and sex were slightly superior to otholith morphological features for age classification purposes. The highest classification accuracy achieved was around 75 per cent.

Grading Fish from Images

Odone *et al.* (2001) have developed a trainable computer vision system for determination of fish weight from image measurements using support vector machine. The system takes image measurements from top and side views to live fish sliding through a transparent channel. After training stage of SVM, the system learns the relation between fish weight and shape parameters. This system was able to grade fish at the rate of three fish per second.

SVM Software

1. *SVMlight*: A popular implementation of the SVM algorithm by Thorsten Joachims; it can be used to solve classification, regression and ranking problems.
2. *SVMProt*: Protein Functional Family Prediction.
3. *LIBSVM*: A Library for Support Vector Machines.
4. *MySVM*: a popular SVM.
5. *LS-SVMLab*: Matlab/C SVM toolbox–well-documented, many features.
6. *Gist*: implementation of the SVM algorithm with feature selection.
7. *SimpleSVM*: SimpleSVM toolbox for Matlab.
8. *The Kernel-Machine Library (GNU)*: C++ template library for Support Vector Machines.

ANN vs. SVM

ANN and SVM both compliment each other. Most of the traditional neural network models seek to minimize the training error by implementing the empirical risk minimization principal, where as the SVMs implement the structural risk minimization principle which attempts to mimimize an upper bound on the generalization error by striking a right balance between the training error and the capacity of machine. The solution of traditional neural network models may tend to fall into a local optimal solution, whereas global optimal solution is guaranteed for SVM (Haykin, 2003).

Conclusions

This chapter serves as an introduction to artificial neural networks and support vector machines with emphasis to biological data analysis and modelling. It presents a brief description of ANNs and SVMs, the concepts, underlying aspects, and its applications. Some popular ANN and SVM architectures and algorithms are also discussed. Guidelines for applying these techniques for data analysis and modelling are presented that may be useful for researches in fisheries/aquaculture.

References

Bermejo, S., 2007. Fish age classification based on length, weight, sex and otolith morphological features. Fisheries Research, 84:270-274

Boser, B. E., Guyon, I. M. and Vapnik, V. N., 1992. A training algorithm for optimal margin classifiers. Proceedings of the fifth annual workshop on Computational learning theory, Pittsburgh, USA, pp.144-152.

Fausett, L., 1994. Fundamentals of neural networks, Architectures, Algorithms and Applications. Prentice Hall, Englewood Cliffs, N.J.

Haykin, S., 2003. Neural networks: a comprehensive foundation. Mac-Millan, New York.

Hebb Donald, O., 1949. The organization of Behaviour, Wiley, NY.

Hyun, K., Song, M., Kim, S., and Chon, T., 2005. Using an artificial neural network to parameterize long-term fisheries data from South Korea. Aquatic Sciences, 63:382-389.

Kohonen, T., 1990. The self organizing map. Proc. IEEE, 78:1464-1480.

Lae, R., Lek, S. and Moreau, J., 1999. Predicting fish yield of African lakes using neural networks. Ecological modelling, 120:325-335.

Lek, S., Belaud, A., Dimopoulos, I., Lauga, J. and Moreau, J., 1995. Improved estimation, using neural networks, of the food consumption of fish populations. Marine and Freshwater Research, 46(8):1229–1236.

Leung, P.S. and Tran, L. T., 2000. Predicting shrimp disease occurrence: artificial neural networks vs. logistic regression. Aquaculture, 187:35-46

McCulloch, W. S. and Pitts, W., 1943. A logic calculus of the ideas immanent in nervous activity. Bull. of Math. Biophys., 5:115-133.

Odone, F., Trucco, E. and Verri, A., 2001. A trainable system for grading fish from images. Applied Artificial Intelligence, 15:735-745.

Rajsekharan, S., and Vijailakshmi Pai, G. A., 2003. Neural Networks, Fuzzy Logic and Genetic Algorithms: Synthesis and Applications. Prentice Hall of India Pvt. Ltd., New Delhi, 439 p.

Rao, V. and Rao, H., 1996. C++ Neural network and fuzzy logic. BPB Publications, New Delhi. 551p.

Rumelhart, D. E., Hinton, G. E. and Williams, R. J., 1986. Learning internal representation by error propagation. Parallel distributed processing, MIT Press, Cambridge, Mak., 1:318-362.

Ruohonen, K., 1999. Modelling fish food consumption with artificial neural networks. Aquaculture research, 30:545.

Tsoukalas Lefteri, H., and Uhrig Robert, H., 1997. Fuzzy and Neural Approaches in Engineering. John Wiley and Sons Inc.

Vapnik, V., 1995. The Nature of Statistical Learning Theory. Springer-Verlag.

Wang, H.C., Dopazo, J., de la Fraga, L. G., Zhu, Y.P. and Carazo, J.M., 1998b. Self–organizing tree-growing network for the classification of protein sequences: Protein Sci, 7:2613-2622.

Zadeh, L.A., 1965. Fuzzy sets. Information and Control, 8: 338-353.

Modelling, Forecasting, Artificial Neural Network and Expert System in Fisheries *Pages 220–233*
Editors: **Ajit Kumar Roy & Niranjan Sarangi**
Published by: **DAYA PUBLISHING HOUSE (2009)**

Chapter 16

An Econometric Model for Productivity Analysis of Aquaculture Farms

A.K. Roy, G.S. Saha, N. Sarangi and Nibedita Jena

Introduction

India's freshwater aquaculture resources includes 2.35 million hectare of ponds and tanks, 1.3 million hectare of oxbow lakes and derelict waters, 0.19 million km of rivers and canals, 3.15 million ha of reservoirs (Anon, 2006 a). Ponds and tanks are the prime resources for aquaculture, however only about 0.8–0.9 million ha is used for aquaculture currently. Indian aquaculture has demonstrated a six and half fold growth over the last two decades with freshwater aquaculture contributing over 95 per cent of the total aquaculture production. Three Indian major carps namely catla (*Catla catla*), rohu (*Labeo rohita*) and mrigal (*C. mrigala*) contribute significantly of the bulk of production of over 1.8 million tones (FAO, 2005). Latest development in this sector is the establishment of National Fisheries Development Board with huge budget provision to improve production, processing, storage, distribution and marketing of fisheries products (Anon, 2006 b). The total aquaculture production of 2.47 million tones that constitute 4.2 per cent of global production was valued at US$ 2.9 billion tones (FAO, 2006).

Aquaculture plays a vital role in ensuring future fish availability for food security and nutrition in India. This sector has to develop and grow in an economically viable and environmentally sustainable fashion. That is possible through improved water management, better feeding strategies, genetic improvement of cultivable species and better health management leading to enhanced production efficiency of aquaculture at farm level. Therefore, measuring Technical Efficiency (TE) at farm level and identifying factors associated with efficient production and assessing potential for and sources of future improvements are need-of-the-hour for establishing sustainable aquaculture productivity. Instead of increasing the use of inputs to increase production, efforts should be made towards output growth through improved TE, which means to produce more by utilizing inputs at hand more efficiently.

In recent years, farmers have intensified their effort towards aquaculture and as a result demand for natural resources like water, land and seed have increased. This phenomenon might have adverse environmental effects which may affect sustainability. Under that situation, enhanced production through improvement of technical efficiency may lead to sustainable production and consistent supply, through efficient use of resources and productivity at farm level.

Need for Productivity Analysis

World Trade Organisation (WTO) in 1995 with 146 member countries including India has made a number of agreements that regulate international trade specifying what countries can do and not do. Most relevant WTO trade rules are (i) Agreement on sanitary and phytosanitary measures (SPS) (ii) Agreement on technical barriers to trade (TBT), (iii) Subsidies agreement (iv) Anti-dumping agreement (v) Agreement on safeguards and (vi) WTO dispute settlement mechanism.

In Doha Development Agenda, fish and fisheries products are not covered by agreement in agriculture *i.e.* more stringent rules for subsidies in fisheries than for agricultural products. Fish was negotiated under market access for nonagricultural products. It was proposed to reclassify, eliminate or reduce capacity-enhancing subsidies. Therefore, it is presumed that due to imposed regulations, trade flow, economic efficiency and welfare may be affected posing a shift competition of exporters in the international market.

To face the international challenges from WTO and GAAT and domestic phenomenon of declining trend of marine fish production emphasis is to be given for efficient management of aquaculture resources. It is generally suggested that instead of increasing the use of inputs to increase production, efforts should be made towards output growth through improved TE that means to produce more by utilizing inputs at hand more efficiently. Role of aquaculture in rural economic development of India is well established. Aquaculture is the fastest growing sector. For sustainable production, rational use of resources is necessary. For rational use of scarce resources, understanding of the pattern of input use and knowledge impact of farm specific variables are necessary. This can be achieved through efficiency measurement of the aquaculture farms spread all over the country, which in turn will increase the efficiency of resource use and productivity at farm level.

Further it is reported that farmers in developing countries fail to exploit fully the potential of a technology making inefficient decision due to various reason of which management capacity is an important one. Various reports are available in literature on potential enhancement of production through efficient use of resources. The profit of rice farmers in Pakistan's Punjab could be increased by 28 per cent through improved efficiency, which is found positively related with education and timeliness of input use (Ali and Flinn, 1989). Average technical efficiency for Nepalese agriculture is 84 per cent for rice production suggesting that there is a potential of 16 per cent increase in the farm output (Belbase and Grabowski, 1985). It is also reported that the "poor but efficient hypothesis" may have overlooked important potential for increasing agricultural productivity (Battese, 1992 and Bravo-Ureta and Pinheiro 1993).

Review of Literature

Scientific studies on statistical and econometric aspects like development of sampling frame, statistics of carp culture, socio economic aspects, economic performance of aquaculture, factor share, database on socioeconomic aspect, farm size aquaculture productivity relationship, break even analysis, marketing channel and technical efficiency study are available (Roy *et al.*, 2002; 2001 a; 2001 b; 2002 b; 2004 a; 2004 b; Sahoo *et al.*, 2000; 2001 a; Saradhi *et al.* 2002; Roy *et al.*, 2006 and Roy *et al.*, 2007). A review of literature reveals that estimation of technical efficiency applying frontier analysis is very limited in aquaculture sector in India (Jayaraman, 1998) and abroad (Sharma and Leung, 1998; Sharma and Leung, 2000 a, b; Dey *et al.* 2000; Iinuma *et al.* 1999). The present chapter is aimed at describing methods for measuring the technical efficiency of aquaculture farms and also to identify the importance of TE determinants demonstrating an empirical example, which in turn may help in drawing policy decisions for the aquaculture sector.

Data Sources for an Empirical Case Study

The present analysis was carried out on a data set of 221 aquaculture farms located at Kolleru Lake area that was collected using multistage stratified random sampling covering 73 revenue villages spread over 9 Mandals of West Godavari and Krishna districts of Andhra Pradesh under the AP-Cess fund project entitled "Technological innovations in aquaculture and its effects on sustainability of farming systems in Andhra Pradesh" funded by Indian Council of Agricultural Research (ICAR) during 1998-2002. Farmers of Kolleru Lake generally practice semi-intensive composite culture of two Indian major carps namely catla (*Catla catla*) and rohu (*Labeo rohita*) at a species ratio of 1:4 and at a stocking density ranging from 5–10 thousand advanced fingerling/ha with higher level of input recommended for composite fish culture adopting single stocking and single harvesting techniques. The earlier practice of five species poly culture of carps (IMC and exotic) is now replaced by a two species (Rohu and Catla) culture system. This technology is the modified version of composite fish culture technology (Chaudhuri *et al.*, 1975) that the farmers of Kolleru Lake area have perfected with respect to species composition and input use, which were driven by growth of cultivable species, market demand, consumer preference, keeping quality and economics. The technology delivers a yield ranging from 7-15 tones/ha/yr.

Input and Output Variables and Sample Characteristics

All input and output variables used in the stochastic frontier production function model in this study are clearly defined along with measurement units and summary statistics of organic manure, inorganic fertilizer, lime, chemicals and drugs, feed applied, types of labor engaged and socio-economic profiles like religion, caste, age, no. of children, education, experience, asset, etc. of the farm operators are reported (Roy *et al.*, 2007).

Survey reveals that the average pond size of our sample was 8.7 ha and the range was 0.8–61 ha. The primary occupation of 79 per cent of the farmers is aquaculture. 96 per cent of the farmers receive water from the adjoining irrigation canals and 4 per cent take water from natural sources. Further it revealed that 83 per cent farmers reported renovating their ponds as part of pond management practices. All farmers cultured catla and rohu at a species ratio of 1:4 and at an average stocking biomass of 1939 kg/ha. Single stocking and single harvesting practices was resorted by 95 per cent of the farmers. 97 per cent of farmers applies feed in pond water using perforated bags hanged from poles and submerged. Average yield of fish was 10.63 t/ha/year.

Econometric Model for Estimation of Technical Efficiency and Test of Hypothesis

The frontier production function defines potential output that can be produced by a farm with a given level of input and technology. It is well known that the level of technical efficiency of a particular farm is characterized by the relationship between observed production and some ideal production (Greene, 1993). Farrell (1957) first developed frontier production models. Since then several approaches to efficiency assessment have been developed that can be classified into two broad categories: parametric and nonparametric. Frontier techniques have been widely used in determining the productive performance of farms which may be estimated either by a stochastic frontier production function (Aigner *et al.*, 1977; Meeusen and van der Broeck 1977) or non-parametric linear programming approach, known as data envelopment analysis (DEA). The stochastic frontier approach is preferred for assessing efficiency in agriculture because of the inherent stochasticity involved (Kirkley *et al.*, 1995; Coelli *et al.*, 1998). There are several reviews of the applications of the stochastic frontier approach in agriculture (Battese and Coelli 1992; Bravo-Ureta and Pinheiro 1993; Coelli 1995). Most applications of frontier analysis in Asian aquaculture have used the stochastic approach (Sharma and Leung, 1998; Sharma, 1999; Bimbao, *et al.*, 2000; Dey *et al.*, 2000; Gunaratne and Leung, 2001; Sharma and Leung, 2000 a; b; and Irz and McKenzie 2003). Inefficiency models may be estimated with either a one step or a two-step process. For the two-step procedure the production frontier is first estimated and the technical efficiency of each farm is derived. These are subsequently regressed against a set of variables, which are hypothesised to influence the farm's efficiency. A problem with the two-stage procedure is the inconsistency in the assumptions about the distribution of the inefficiencies. In the first stage, the inefficiencies are assumed to be independently and identically distributed is order to estimate their values. However, in the second stage, the estimated inefficiencies distributions are assumed to be a function of number of farm specific factors, and hence are not identically distributed unless all the coefficients of the factors are simultaneously equal to zero (Coelli, Rao and Battese, 1998). FRONTIER uses the ideas of Kumbhakar, Ghosh and McKenzie (1991) and Reifschneider and Stevenson (1991) and estimates all of the parameters in one step to overcome this inconsistency. The inefficiency effects are defined as functions of farm specific factors (as in the two stage approach) but they are then incorporated directly into the maximum likelihood estimation.

The stochastic production frontier for sample carp culture ponds is specified as follows taking into account the model developed and proposed by Aigner *et al.* (1977) and Meeusen and van der Broeck (1977).

$$LnY_i = \beta_0 + \sum_{k=1}^{8} \beta_k LnX_{ki} + V_i - U_i \qquad (1)$$

where, subscript i refer to the i-th farm in the sample; Ln represents the natural logarithm; Y is output variable and Xs are input variables. βs are unknown parameters to be estimated; Vi is an independently and identically distributed N $(0, \sigma_v^2)$ random error; and the Ui is a non-negative random variable associated with technical inefficiency in production, which is assumed to be independently and identically distributed and the inefficiency is assumed to here a truncated normal distribution with mean, μi and variance, σ_v^2 $(|N(\mu_i, \sigma_v^2)|)$. According to Battese and Coelli (1995), the technical inefficiency distribution parameter, μ_i is defined as:

$$\mu_i = \delta_0 + \sum_{m=1}^{25} \delta_m Z_{mi} \qquad (2)$$

where, Zs are 25 farm-specific variables and δs are unknown parameters to be estimated. Since the dependent variable in (2) is defined in terms of technical inefficiency, a farm-specific variable associated with a negative coefficient will have a positive impact on technical efficiency (Sharma and Leung, 2000a).

Socio economic and demographic variables are reported to have some impact on technical efficiency. Similarly farm specific variables are also expected to have an impact over technical efficiency affecting carp yield. Farm specific technical efficiency of the i-th sample farm (TEi) is obtained using the relationship.

Antilog and Expected Value

$$TEi = \exp(-Ui) \qquad (3)$$

The prediction of technical efficiencies is based on the conditional expectation of expression in (3) given the model of specifications (Battese and Coelli, 1988). The parameters for the stochastic production frontier model in (1) and those for the technical inefficiency model in (2) were estimated simultaneously with maximum-likelihood estimation (MLE) using an econometric computer program, FRONTIER 4.1 (Coelli, 1994), that estimates the variance of the likelihood function as:

$$\sigma^2 = \sigma_u^2 + \sigma_v^2 \text{ and}$$

$$\gamma = \sigma_u^2/\sigma^2$$

σ^2 and r are not respondent estimates $\gamma = 1 - \sigma u^2/\sigma^2$

Hypotheses Tests of

The following hypotheses are required to be tested with a generalized likelihood ratio test to ensure that inefficiency effects were absent from the model.

H_0: $\mu = 0$, the null hypothesis specifies that a simpler half normal distribution is an adequate distribution of data.

H_0: $\gamma = \delta_0 = \delta_1 = \ldots\ldots = \delta_{25} = 0$, technical inefficiency does not exist.

H_0: $\gamma = 0$, technical inefficiency is deterministic. Hence explanatory variables in (2) must be included in along with input and other relevant variables.

H_0: $\delta_1 = \delta_2 = \ldots\ldots = \delta_{25} = 0$, technical inefficiency effects follows a standard truncated-normal distribution with no technical inefficiency effects.

The null hypothesis can be tested using the generalized likelihood –ratio statistic, λ, given by:

$$\lambda = -2\,[\mathrm{Ln}\,\{L\,(H_0)\} - \mathrm{Ln}\,\{L\,(H_1)\}] \tag{4}$$

where, $L\,(H_0)$ and $L\,(H_1)$ denote the values of likelihood function under the null (H_0) and alternative (H_1) hypotheses respectively. If the given null hypothesis is true, λ has approximately χ^2 distribution or mixed χ^2 distribution under $\gamma = 0$ (Coelli, 1995).

Analytical Output

Maximum Livelihood Estimates and Test of Hypotheses

The hypothesis that technical inefficiency effects have a half-normal distribution with mean zero was rejected. Likewise the hypothesis that there is no technical inefficiency was also rejected. The hypothesis that technical inefficiency is deterministic is rejected. The null hypothesis that all the parameters in the technical inefficiency model except the constant term are zero *i.e.*, the technical inefficiency effects have the same truncated normal distribution was also rejected (Table 16.1). The likelihood ratio tests indicate that technical inefficiency effects are significant in explaining the variation in production of the carp farms located at Kolleru Lake.

Frontier Functional Analysis

Qualitative and Quantitative data analysis was carried out for estimation of Technical Efficiency (TE) and determinants of Technical Inefficiency of Aquaculture farms of Kolleru Lake, Andhra Pradesh, India (Roy *et al.*, 2007). The frontier production function that was used in analysis defines potential output that can be produced by a farm with given level of inputs and technology. In the first stage, farm-specific technical efficiency scores are estimated using stochastic frontier production function (Aigner, 1977). In the second stage, the influence of socioeconomic and farm specific variables are examined using an established model proposed by Battese and Coelli, 1995. The analysis was carried out using software FRONTIER. Various types of variables considered for the study are:

Table 16.1: Generalised Hypothesis Tests for Model Specification and Statistical Assumptions

Null Hypothesis (H_0)	*Log (Likelihood)*	*Test Statistic (λ)*	*Critical Value ($\chi^2_{0.05}$)*	*Decision*
$\mu = 0$	210.79915	7.92248	3.84	Reject H_0
$\gamma = \delta_0 = \delta_1 = \ldots\ldots = \delta_{25} = 0$	158.77339	111.974	39.53[a]	Reject H_0
$\gamma = 0$	198.16121	33.19836	3.84	Reject H_0
$\delta_1 = \delta_2 \ldots\ldots = \delta_{25} = 0$	183.23366	63.05346	37.63	Reject H_0

a: $\chi^2_{0.05}$ is obtained from Kodde and Plam (1986, P.1246).

Output

Yield of Fish

Input

Organic manure, Inorganic fertilizer, Lime, Stocking weight, Chemicals and drugs, Feed, Labour and Water Management Cost.

Socio-economic and Farm Specific Variables

Religion, Caste, Age, Number of children in the family, Education, Experience in carp culture, Primary occupation, Pond size. Pond renovation, Asset cost, Ownership type, Source of water, Depth of water, Source of seed, Sources of advice taken, Loan, Harvesting technique, Method of application of drug, Methods of application of feed, Calamities faced, Periodical netting for biomass checking, Types of labour, Types of organic manure, Types of feed and Disease encountered.

Interpretation of Input-Output Analysis

From Table 16.2 it is apparent the elasticity of output for feed was highest (0.2001) (P<001) followed by organic manure (0.1411) (P<001). This is indicative of the fact that both feed and organic manure have major influence on yield. Inorganic fertilizer, stocking weight, labour and lime were not found statistically significant. Both chemicals and drugs and water management coefficient showed no impact over yield. The estimated values of s_u^2 and s_v^2 were found to be 0.008408791 and 0.000006594 respectively indicating that technical inefficiency s_u^2 is greater than the inefficiency caused by stochastic factor s_v^2. These values indicate that the differences between the observed (actual) and frontier (potential) output are due to inefficiency and not chance alone. A value of one indicate the difference attributed to the farmers' less than efficient use of technology *i.e.*, technical inefficiency (Coelli, 1995). In the present case the estimated value of γ, which is the ratio of variance of farm-specific TE to total variance of output, is 0.9992 which suggest that 99.92 per cent of error variation in output among the farms is due to technical inefficiency. This indicates the fact that the random component of inefficiency does not have a significant contribution in the analysis. This result is identical to the observations of Sharma *et al.* 1997, Hjalmarsson *et al.* 1996, Coelli and Battese, 1996, Kalirajan, 1981, Ajibefun *et al.* 1996, Ali and Flinn, 1989 suggesting that technical inefficiency is important in explaining the levels and variations in carp yield in Kolleru Lake. The estimated value of λ was found be 35.71020897 which is positive and statistically significant, confirming an effect of TE.

Table 16.2: Maximum-likelihood Estimates of Stochastic Production Frontier and Technical Inefficiency Models for Carp Production in Kolleru Lake, Andhra Pradesh, India

	Parameter	*Coefficient*	*Standard–error*	*T–ratio*
Stochastic production frontier model				
Constant	β_0	1.2374793***	0.19531434	6.3358343
Organic manure (X_1)	β_1	0.14112453***	0.037727305	3.7406470
Inorganic fertilizer (X_2)	β_2	0.024032115	0.023807879	1.0094186
Lime (X_3)	β_3	0.033957713	0.026476323	1.2825691
Stocking weight (X_4)	β_4	0.026828252*	0.019799683	1.3549839
Chemicals and Drugs (X_5)	β_5	–0.014569666	0.017009698	–0.85655054
Feed (X_6)	β_6	0.20010391***	0.036975060	5.4118616
Labor (X_7)	β_7	0.011732542	0.011203558	1.0472157
Water management (X8)	β_8	–0.022573018*	0.011707627	–1.9280609
Technical inefficiency model				
Constant	δ_0	0.55002287	0.60348210	0.911415
Religion (Z_1)	δ_1	–0.085277684*	0.050847024	–1.6771421
Caste (Z_2)	δ_2	0.013630104*	0.0097874157	1.3926152
Age (Z_3)	δ_3	–0.00051804104	0.0013822222	–0.3747885
No of children (Z_4)	δ_4	0.0041073793	0.0084882335	0.4838909
Education (Z_5)	δ_5	–0.0016462576	0.0040847498	–0.4030253
Experience (Z_6)	δ_6	0.00022949786	0.0022022899	0.10420874
Primary occupation (Z_7)	δ_7	–0.027911207*	0.022389016	–1.2466473
Pond size (Z_8)	δ_8	–0.0026728321**	0.0011043401	–2.4202980
Renovation (Z_9)	δ_9	0.023636240	0.019656041	1.2024924
Asset cost (Z_{10})	δ_{10}	0.00000041221	0.0000010251	0.4020827
Ownership type (Z_{11})	δ_{11}	–0.037148235	0.033218818	–1.1182889
Source of water (Z_{12})	δ_{12}	–0.097534707**	0.040034542	–2.4362639
Depth of water (Z_{13})	δ_{13}	–0.0008716901	0.023454639	–0.0371649

Contd...

Table 16.2–Contd...

	Parameter	Coefficient	Standard–error	T–ratio
Source of seed (Z_{14})	δ_{14}	0.030013137*	0.016411887	1.8287439
Sources of Advice taken (Z_{15})	δ_{15}	–0.0045832209	0.010281450	–0.4457757
Loan (Z_{16})	δ_{16}	0.021655156	0.024542989	0.8823357
Harvesting technique (Z_{17})	δ_{17}	0.059316720*	0.040854522	1.4519010
Method of application of drug (Z_{18})	δ_{18}	–0.015432842*	0.010162719	–1.5185742
Method of application of feed (Z_{19})	δ_{19}	0.0033167997	0.043048858	0.0770473
Calamities faced (Z_{20})	δ_{20}	–0.029885695*	0.018107579	–1.6504523
Periodical netting for biomass checking (Z_{21})	δ_{21}	0.068515871**	0.025246614	2.7138638
Types of labour (Z_{22})	δ_{22}	–0.042361853*	0.030499051	–1.3889565
Type of organic manure (Z_{23})	δ_{23}	0.0091244716	0.013217207	0.6903479
Type of feed (Z24)	δ_{24}	–0.019243984	0.016299137	–1.1806750
Disease encountered (Z_{25})	δ_{25}	–0.00026916853	0.0040550140	–0.0663799
	σ^2	0.0084153852	0.00077490850	10.8598440
	γ	0.99921654	1.3397727	0.74581048
	Log likelihood function	214.76039		
	LR test	63.053464		
	Mean TE	0.72608099		

Source: Roy *et al.*, 2006a.

***, **, * Significant at 1, 5 and 10 percent level of significance respectively.

Estimation of Technical Efficiency

Table 16.3: Estimated Technical Efficacy Score of Aquaculture Farms of Kolleru Lake

Efficiency Score Class	*Number of Farmers*	*% of farmer*	*Cumulative Frequency of Farmer*
0.50–0.60	6	2.71	17
0.60–0.70	90	40.72	125
0.70–0.80	88	39.82	199
0.80–0.90	30	13.58	218
0.90–1.0	7	3.17	221
Total	221	100.00	—

Source: Roy *et al.*, 2006a.

Interpretations on TE Score

Technical efficiency (TE) scores ranged from 0.5051 to 0.9899 with a mean of 0.7260 and modal class is 0.6 to 0.7. In the present study it revealed that over 56 per cent of the farmers operate with efficiency score greater than 0.7 and no farmer operate below the technical efficiency of 0.50.

Potential for Technical Efficiency Improvement

Analytical results suggest that on an average the farmers of Kolleru Lake area are able to obtain about 72.60 per cent of potential output from the given set of production inputs. Therefore, in the short run there is a scope of increasing the productivity of carps by around 27.40 per cent by adopting the technology and techniques used by efficient farms. Further, it is observed that if the average farmer in the sample was to achieve the TE level of its most efficient farmer counterpart then the average farmer could realize 26.7 per cent cost saving (1–[72.60/98.99]). Similarly the most technically inefficient farmer may achieve cost saving of 48.9 per cent (*i.e.*, 1–[50.51/98.99]).

Salient Findings of Input-Output Analysis

Organic manure, inorganic fertilizer, lime, stocking weight, feed, and labor demonstrated positive impact on fish production whereas chemicals drugs and water management cost have negative impact on fish production. However, maximum elasticity of output was demonstrated by feed (0.2001) followed by organic manure (0.1411) that are found statistically significant justifying the significant relatively better impact of these variables in fish production.

Determinants of Technical Inefficiency

Socio-economic and farm specific variables like religion, age, education, primary occupation, pond size, ownership type, source of water, depth of water, sources of advice taken, method of application of drugs, calamities faced, types of labour, types of feed and disease encountered showed (+ve) impact on TE. But out of all these only pond size and source of water showed significant (+ve) impact on TE ($p<0.05$). Types of labour, religion, method of application of drug, primary occupation and calamities showed significant (+ve) impact ($p<0.10$) on TE. On the other hand caste, number of children, experience, renovation, asset cost, source of seed, loan, harvesting technique, method of application of feed, periodical netting for biomass checking and types of organic manure showed (-ve) impact on TE. Among all these only periodical netting for biomass checking showed significant (-ve) impact ($p<0.01$) and source of seed, harvesting technique and caste showed significant (-ve) impact ($p<0.10$).

Policy Implications

The stochastic frontier estimate demonstrates the average technical efficiency of 72.60 per cent. This indicates that there is scope for further increasing the output by 27.40 per cent without increasing the levels of inputs. The use of stochastic frontier analysis enabled ranking of the carp culture farms based on efficiency of yield of fish facilitating identification of the most efficient and inefficient farms across various carp farming systems operating in Kolleru lake area. The efficient farms may serve as a model farm for improving the efficiency of carp production in Kolleru lake area as well as throughout the country. Research on other productive efficiencies especially allocative and economic efficiences may the carried out.

Reference

Aigner, D.J., Lovell, C.A.K., Schmidt, P.J., 1977. Formulation and estimation of stochastic frontier production function models. *J. Econometrics* 6, 21-37.

Ali, M., Chaudhury, M. A., 1990. Inter-regional farm efficiency in Pakistan's Punjab: A frontier production function study. *Journal of Agricultural Economics*, 41(1), January 62-74.

Anon, 2006a. Pamphlet, National Fisheries Development Board. Department of Animal Husbandry, Dairying and Fisheries, Ministry of Agriculture, Govt. of India, Block 401-402, Maitri Vihar, Huda Commercial Complex, Ameerpet, Hyderabad-500038, Andhra Pradesh, India.

Anon, 2006b. Inaugural address of Sri Sharad Pawar, Hon'ble Union Minister for Agriculture, Consumer Affairs, Food and Public Distribution. 9th September 2006, National Fisheries Development Board, Hyderabad, India.

Ajibefun, I.A., Battese, G.E., Daramola, A.G., 1996. Investigation of factors influencing the technical efficiencies of smallholder croppers in Nigeria, CEPA working papers, No. 10/96, Department of Econometrics, University of England, Armidale, 19.

Ali, M., Flinn, J.C., 1989. Profit efficiency among basmati rice producers in Pakistan Punjab. *American Journal of Agricultural Economics*, 71, 303-310

Battese, G.E., Coelli, T.J., 1992. Frontier production function, technical efficiency and panel data: with application to paddy farmers in India. *J. Productiv. Anal.*, 3, 153-169.

Battese, G.E., Coelli, T.J., 1988. Prediction of farm-level technical efficiencies with a generalised frontier production function and panel data. *Journal of Econometrics*, 38, 387-399.

Battese, G.E., Coelli, T.J., 1995. A model for technical inefficiency effects in a stochastic frontier function for panel data. *Empirical Economics* 20, 325-332.

Bimbao, G. B., Paraguas, F. J., Dey, M. M., Eknath, A. E., 2000. Socio-economics and production efficiency of Tilapia Hatchery Operations in Philippines. *Aquaculture Economics and Management*, 4(1) and (2), 31-46.

Bravo-Ureta, E., Boris, E., Antonio, Pinheiro, E., 1993. Efficiency analysis of developing country agriculture: A review of the frontier function literature. *Agricultural and Resource Economics Review* 22(1), 88-101.

Chaudhuri, H., Chakrabarty, R. D., Sen, P. R., Rao, N. G. S., Jena, S., 1975. A new high in fish production in India with record yields by composite fish culture in freshwater ponds. *Aquaculture*, 6(1975), 343-355.

Coelli, T.J., 1994. A guide to FRONTIER version 4.1: a computer programme for stochastic frontier production and cost function estimation. Department of Econometrics, Univ. of New England, Australia.

Coelli, T.J., 1995. Estimators and hypothesis tests for stochastic frontier function: a Monte Carlo analysis. *Journal of Productivity Analysis* 6, 247-268

Coelli, T.J., Rao, D.S.P., Battese, G.E., 1998. *An introduction to efficiency and productivity analysis.* USA: Kluwer Academic Publishers, USA.

Coelli, T. J., Battese, G.E., 1996. Identification of factors which influence the technical inefficiency of Indian Farmers. *Australian Journal of Agricultural Economics,* 40(2), 103-128.

Dey, M. M., Paraguas, F. J., Bimbao, G.B., Regaspi, P.B., 2000. Technical efficiency of Tilapia grow out pond operations in the Philippines. *Aquaculture Economics and Management,* 4(1) and (2), 33-47.

FAO, 2005. Aquaculture production, 2003. Yearbook of fishery statistics–Vol.96/2. *Food and Agriculture Organization of the United Nations,* Rome, Italy.

FAO, 2006. State of World Aquaculture: 2006. *FAO Fisheries Technical Paper.* FAO. Rome.

Farrell, M.J., 1957. The measurement of productive efficiency. *Journal of the Royal Statistical Society,* series A120, Part 3, 253-281.

Greene, W.H., 1993. The econometric approach to efficiency analysis, *In*: Fried, H.O., Lovell, C.A.K. and Schmidt, S.S. (Eds), *The Measurement of Productive Efficiency,* Oxford University Press, New York, 68-119.

Gunaratne, L.H.P., Leung, P. S., 2001. Asian Black Tiger Shrimp Industry: A meta-production frontier analysis, *In*: P.S. Leung, K. R. Sharma (Eds.). *Economics and Management of Shrimp and Carp Farming in Asia,* Network of Aquaculture Centres in Asia-Pacific, Bangkok, Thailand.

Hjalmarsson, L., Kumbhakar, S.C., Heshmati, A., 1996. DEA, DFA and SFA: A Comparison. *Journal of Productivity Analysis,* 7,303-327.

Iinuma, M., Sharma, K. R., Leung, P. S., 1999. Technical efficiency of carp pond culture in peninsula Malaysia: an application of stochastic production frontier and technical inefficiency model. *Aquaculture* 175, 199-213..

Irz, X., McKenzie, V., 2003. Profitability and technical efficiency of aquaculture systems in Pampanga, Philippines. *Aquaculture Economics and Management,* 7(3) and (4), 195-211.

Jayaraman, R., 1998. Economics and technical efficiency in carp culture in Thanjavur District, Tamil Nadu State, India. *In*: Eide, A., Vassdal, T. (Eds.), *Proceedings of 9th International Conference of the International Institute of Fisheries Economics and Trade,* Tromso, July 8-11, 1998. Norwegian College of Fishery Science, Tromso, Norway, 71-82.

Kalirajan, K., 1981. An econometric analysis of yield variability in paddy production. *Canadian Journal of Agricultural Economics,* 29, 167-80.

Kirkley, J.E., Squires, D.E., Strand, I.E., 1995. Assessing technical efficiency in commercial fisheries: The Mid-Atlantic Sea Scallop Fishery. *Am. J. Agric. Econ.,* 77(3), 686-97.

Kodde, D.A. and Palm, F.C., 1986. Wald criteria for jointly equality and inequality restrictions. *Econometrica* 54, 1243-1248.

Kumbhakar, S.C. and Bhattacharya, A. (1992). Price distortion and resource use efficiency in Indian agriculture: a restricted profit function approach. *Review of Economics and Statistics*, 74, 231-239.

Kumbhakar, S.C., Ghosh, S. and Mcguckin, J., 1991. A generalized production frontier approach for estimating determinants of inefficiency in US dairy farms. *J. Bus. Econ. Statist.* 9, 279-286.

Meeusen, W. and Van der Broeck, J., 1977. Efficiency Estimation from Cobb-Douglas Production Functions with Composed Errors. *Int. Econ. Rev.*, 18: 435-444.

Reifschneider, D. and Stevenson, R., 1991. Systematic departures from the frontier: A framework for the analysis of firm inefficiency. *International Economic Review*, 32, 715-723.

Roy, A.K., Saradhi, K.P., Sahoo, K.N. and Saha, G.S., 2002. Statistics of carp culture in Eluru and Kaikalur Mandals of Kolleru Lake in Andhra Pradesh, India. *In*: Ayyappan, S., Jena, J. K., Mohan Joseph, M. (Eds.), *The Fifth Indian Fisheries Forum Proceedings AFSIB*, Mangalore and AoA, Bhubaneswar, India, 409-411.

Roy A. K., Sahoo, K. N., Saradhi, K.P. and Saha, G.S., 2001a. Aquaculture in Kolleru Lake area for socio-economic development. *Environment and Ecology*, 19(4), 927-931.

Roy, A.K., Sahoo, K.N., Saradhi, K.P. and Saha G.S., 2001b. A socio economic study on aqua culturists of Eluru and Kaikalur Mandals of Kolleru Lake area of Andhra Pradesh. *In*: Sinha, M., Kumar, Dhirendra, Katiha, P.K. (Eds.), *National conference on eco-friendly management of resources for doubling fish production strategies for 21 century held at Central Inland Capture Fisheries Research Institute* (CICFRI), Barrackpore, during 22-23 December, 1999., 176-179

Roy, A.K., Sahoo, K.N., Saradhi, K.P. and Saha, G.S., 2002b. Farm size and aquaculture productivity relationship. *Asian Fisheries Science* 15(2000), 129-134, Asian Fisheries Society, Manila, Philippines.

Roy, A. K., Saradhi, K. P., Saha, G.S., Sahoo K.N. and Antony A., 2004a. CD-ROM Database on socio-economic aspects of carp culture in Kolleru lake. *In*: Roy, A.K., A. Antony and M. Rout (Eds.), *Bioinformatics and Statistics in Fisheries Research*, Central Institute of Freshwater Aquaculture, Kausalyaganga, Bhubneswar, India, 119-123.

Roy, A. K., Saradhi K. P., Sahoo K. N. and Saha G. S., 2004b. Development of sampling frame of cultivable water area in Kolleru Lake of Andhra Pradesh. *In*: Roy, A.K., A. Antony and M. Rout (Eds.), *Bioinformatics and Statistics in Fisheries Research*, Central Institute of Freshwater Aquaculture, Kausalyaganga, Bhubaneswar, India, 353-356.

Roy, A. K., Saha, G. S. and Sarangi, N., 2006. Socio-economic, educational and demographic profile of fish farmers of Kolleru Lake, Andhra Pradesh. *In*: *Compendium of lectures on National Workshop Cum Training on Bioinformatics and*

Statistics in Aquaculture Research, held during 7–10 February, 2006, Central Institute of Freshwater Aquaculture, Kausalyaganga, Bhubaneswar-751002, India.

Roy, A K., Saha, G. S., Sarangi, N. and Jena, Nibedita, 2007. Methods of Estimation of Technical Efficiency of Aquaculture Farms. *In*: Roy, A.K., and N. Sarangi (Eds.), *Bioinformatics and Statistics in Fisheries Research* (Vol. III). Central Institute of Freshwater Aquaculture, Kausalyaganga, Bhubaneswar, India, 439-458.

Sahoo, K.N., Roy A.K., Saradhi, K.P. and Saha, G.S., 2000. Break even analysis of carp culture practice in Eluru Mandal–A Case Study, *In*: *Proc. Fifth Indian Fisheries Forum, held at Central Institute of Freshwater Aquaculture*, Bhubaneswar during 17 –20 Jan 2000, pp. 146.

Sahoo, K. N., Roy, A. K., Saradhi, K. P. and Saha, G. S., 2001a. A study on marketing channels of fish produce of Kolleru Lake. *Journal of Fisheries Economics and Development*, 4(1), 35-44.

Saradhi, K.P., Roy A.K., Sahoo, K.N. and Saha G.S., 2002. On socio-economic database management in Aquaculture. *In*: Ayyappan, S., Jena, J. K., Mohan Joseph, M. (Eds.), *The Fifth Asian Fisheries Forum. Proceedings* AFSIB, Mangalore and AoA, Bhubaneswar, India, 413-415.

Sarangi, N., Kumaraiah, P., Rangacharyulu, P.V. and Giri Bandla, S., 2004. Status of freshwater aquaculture in Krishna–Godavari Delta–A Profile. Central Institute of Freshwater Aquaculture, Kausalyaganga, Bhubaneswar-751002, pp 29.

Sharma, K.R., Leung, P. and Zaleski, H.M., 1997. Productive efficiency of the swine industry in Hawaii: stochastic frontier vs. data envelopment analysis, *Journal of Productivity Analysis*, 8,447-459.

Sharma, K. R. and Leung, P. S., 1998. Technical efficiency of carp production in Nepal: An application of Stochastic Frontier Production Function Approach. *Aquaculture Economics and Management*. Vol.2, No. 3, pp. 129-140.

Sharma, K.R. and Leung, P., 1999. Technical efficiency of the longline fishery in Hawaii: an application of a stochastic production frontier. *Mar. Resource Econ.*, 13, 259-274.

Sharma, K. R., 1999. Technical efficiency of carp production in Pakistan. *Aquaculture Economics and Management*, Vol.3, No. 2, 131-141.

Sharma, K. R. and Leung, P. S., 2000 a. Technical efficiency of carp pond culture in South Asia: An Application of a stochastic meta-production frontier model. *Aquaculture Economics and Management*. Vol. 4, Nos. 3 and 4, 169-189.

Sharma, K.R. and Leung, P. S., 2000 b. Technical efficiency of carp production in India: A stochastic frontier production function analysis. *Aquaculture Research*, 31: 937-958.

Index